Quantum Zero-Error Information Theory

Elloá B. Guedes • Francisco Marcos de Assis
Rex A.C. Medeiros

Quantum Zero-Error
Information Theory

Elloá B. Guedes
Superior School of Technology
Amazonas State University
Manaus, AM, Brazil

Rex A.C. Medeiros
School of Science and Technology
Rio Grande do Norte Federal University
Natal, RN, Brazil

Francisco Marcos de Assis
Center of Electric Engineering
  and Informatics
Campina Grande Federal University
Campina Grande, PB, Brazil

ISBN 978-3-319-82664-6   ISBN 978-3-319-42794-2   (eBook)
DOI 10.1007/978-3-319-42794-2

Printed on acid-free paper

This Springer imprint is published by Springer Nature
The registered company is Springer International Publishing AG Switzerland

"From error to error,
  one discovers the entire truth."

– Sigmund Freud (1856–1939)

Foreword

The field of Quantum Information Science (QIS) has the potential to lead to revolutionary advances in our capability to communicate and process information. It has grown and diversified with a spectacular dynamism since the 1980s, following the work of pioneers such as Charles Bennett, Richard Feynman, or Artur Ekert. Those progresses were rooted in a vision that has proved to be extremely fertile: asking—and answering—questions about information processing by combining quantum physics and computer science approaches. QIS is entering a new and exciting stage of its development, with the emergence of a focused effort on quantum engineering, stimulated by our ability to fabricate increasingly efficient and reliable devices for quantum information processing. This fact may soon allow to experimentally observe the so-called quantum supremacy, i.e., the ability of quantum processing machines to solve information processing tasks that are beyond the reach of existing classical computing devices.

Many important factors can be accounted for to explain these impressive developments, starting of course by the level of maturity reached by classical information processing, computer science, and quantum physics at the end of the twentieth century. We believe that it is particularly interesting, as an introduction to this book on quantum zero-error information theory, to also mention a somehow distinctive feature of QIS: the development of a common language, *quantum information theory*, shared by mathematicians, physicists, and now engineers to talk about information from a quantum perspective.

Standing on the shoulders of Shannon information theory, quantum information theory has gradually become an extremely rich and powerful language, allowing to address not only the fundamental issues of quantum information processing and quantum communications but also the security of cryptographic tasks in a quantum context and, even more recently, quantum thermodynamics.

The present textbook, written by Elloá B. Guedes, Rex A.C. Medeiros, and Francisco Marcos de Assis, is presenting wonderful contributions to quantum information theory, several of them by the authors themselves that have already contributed to making it an ever-richer language. The authors have in particular played a decisive role in this research by introducing, almost 15 years ago, a question that was very original at that time: how could the theory of zero-error information theory, developed in the seminal work of Claude E. Shannon in 1956, be translated and actually extended in a quantum context?

This question has proved to be an extremely fertile topic of research since then, leading to important progress in our understanding of structural properties of channels in quantum information theory as well as fascinating connections with complexity theory. This textbook contains a pedagogical introduction to quantum information theory, offering an approach to this common language that will be particularly well suited to readers with a background in mathematics and information theory. The book also proposes a thorough and original survey of the different important contributions to quantum zero-error information theory, including recent results. We are convinced that this book will be a highly valuable resource for researchers in quantum information and a stimulation for future progresses in zero-error quantum information theory.

Paris, France   Romain Alléaume
May 2016

Preface

This is a textbook on *Quantum Zero-Error Information Theory*. The reader will find an approachable introduction to this subject, from the building blocks of this theory to the latest contributions to the literature. The contents of several original research papers are introduced in a pedagogical way, making it easier for the reader to learn the concepts and depict the developments.

Our dedication to this subject started almost 15 years ago when we were with the IQuanta—Institute for Studies in Quantum Computation and Information, Campina Grande Federal University, Paraíba, Brazil. This institute was founded considering a multidisciplinary body of research having strong interest in learning and contributing to this amazing field of quantum computation and information. Since the definition of the zero-error capacity of quantum channels in 2005, many interesting contributions to this area were given by some of the most important researches in quantum information around the world. Last year, we decided to write a book that contemplate our seminal contributions, as well as the state of the art in this field.

This book starts with an introduction to the fundamentals of quantum information processing in Chap. 2. The main goal of this chapter is to introduce the basic concepts of quantum mechanics for the reader who is not familiar with them. We show how information is represented, processed, and measured in a quantum domain. We also introduce some of the amazing features of the quantum mechanics theory, such as superposition, parallelism, and entanglement. All the concepts presented in this chapter are essential for understanding further results in quantum information theory.

Chapter 3 revisits elementary concepts of classical and quantum information theory. Initially, we give a mathematical definition of information and define the main measures of information, such as entropy and mutual information. We characterize information sources and communication channels in order to enunciate two important theorems proposed by Shannon—the source coding and channel coding theorems. Concerning the quantum information theory, we explore the quantum counterpart of classical concepts already presented. Initially, we define what are quantum states and how information can be encoded within them. Then, we introduce measures of quantum information in terms of the von Neumann entropy. Quantum channels are defined, as well as the accessible information of quantum source. These initial concepts, together with basic results such as the Holevo bound, are sufficient to introduce a variety of quantum channels capacity, such as the $C_{1,1}$ capacity, the HSW capacity, the quantum capacity, and the entanglement-assisted capacity. These capacities illustrate how quantum channels can be used to convey information in many different ways.

Once these background skills are established, Chap. 4 gives an overview of the classical zero-error information theory. The chapter examines the results of an important paper written by Claude E. Shannon in 1956, where he demonstrated how classical channels could be used to transmit information in a scenario where no errors are permitted, instead of allowing an asymptotically small probability of error. We show the characterization of such capacity, its relation with graph theory, the Lovász theta function, and the zero-error capacity of sums and products of classical channels.

Chapter 5 brings the generalization of zero-error capacity to the quantum scenario. Many of the results presented in this chapter, including the definition of the zero-error capacity of quantum channels, were taken from the PhD thesis of Rex A.C. Medeiros, when he was advised by Francisco Marcos de Assis of Campina Grande Federal University and Gérard Cohen and Romain Alléaume of Télécom ParisTech. The chapter also presents contributions from other researchers; the most impressive of which is the superactivation of the zero-error capacity, a phenomenon that has no counterpart in classical theory.

The next two chapters present contributions to the quantum zero-error information theory developed by Elloá B. Guedes during her doctorate under the advisory of Francisco Marcos de Assis at Campina Grande Federal University. Chapter 6 introduces the notion of zero-error secrecy capacity of quantum channels which puts together both zero-error and secrecy capacities. This is a particular scenario where information can be conveyed not only without errors but also in perfect secrecy among two parties sharing a particular class of quantum channels. The chapter starts with some background concepts regarding decoherence-free subspaces and quantum secrecy capacity. Also, we present a characterization of the quantum zero-error secrecy capacity in terms of graphs. Finally, a security analysis of the proposed protocol is made; some examples and related literature are also included.

Chapter 7 introduces a measure of information of quantum source that is based on the zero-error capacity of classical channels, the zero-error accessible information of quantum sources. It has no classical counterpart and measures the maximum amount of information that can be retrieved from a quantum source after a measurement without decoding errors. Besides introducing this concept, the chapter includes some examples and discusses some related works in the literature.

Finally Chap. 8 includes most of the recent contributions to the literature of zero-error information theory. After some discussion about classical and quantum correlations on the proof of Bell's inequality, Gleason and Kochen-Specker theorems are introduced in order to give a formal definition of the quantum chromatic number. A quantum version of the Wielandt's inequality is discussed, followed by the definition of the entangled assisted zero-error capacity of a quantum channel and a very interesting generalization of the Lovász ϑ functional. The quantum clique problem is defined, and we show that the problem of finding the quantum clique belongs to the $\mathcal{QMA}$-complete complexity class. Some of these topics are currently being actively researched and have strong impact on the development of new results in quantum zero-error information theory.

We recommend our readers to follow the dependency diagram below in order to see how the chapters of the book are interconnected. The most important dependencies are emphasized with a continuous line, while suggested reading is shown with a dotted line.

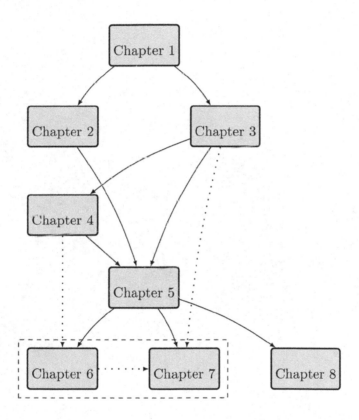

By reading this book, we believe that the reader will follow the path along the most up-to-date developments of the quantum zero-error information theory. To make this task easier, the reader can find suggestions for further reading at the end of each chapter. The references lead to original works where the reader can retrieve the seminal ideas behind every subject discussed. Within the chapters, the readers can also find many detailed examples to illustrate and help in understanding the multiple concepts presented.

Corrections to this edition and suggestions or comments regarding the topics discussed can be sent to our e-mails (ebgcosta@uea.edu.br, rexmedeiros@ect.ufrn.br, fmarcos@dee.ufcg.edu.br). We would be very glad to hear from you!

Manaus, Brazil   Elloá B. Guedes
Natal, Brazil   Rex A.C. Medeiros
Campina Grande, Brazil   Francisco Marcos de Assis

Acknowledgments

We are very grateful to the many people that helped this book come true. We would like to thank our colleagues from the Institute for Studies in Quantum Computation and Information (IQuanta) at Campina Grande Federal University. In particular, we would like to thank Aércio F. de Lima and Bernardo Lula Jr. for their encouragement in our first steps in the quantum world. We had a lot of encouragement from some professors and researchers at Campina Grande Federal University that also helped IQuanta to come true. We would like to thank Bruno Albert, Edmar Candeia, Herman Gomes, Joseana Fechine, and Raimundo Freire. Part of Rex's work on the zero-error capacity of quantum channels was completed when Rex was a graduate student at Télécom ParisTech, France. Rex and Francisco gratefully thank Gérard Cohen, Romain Alléaume, Hugues Randriam, and all folks at Télécom ParisTech Network and Computer Science Department.

The Brazilian research community on quantum computation and information through the Workshop-School of Quantum Information and Computation (WECIQ), held since 2006, always kept our horizons ahead, stimulating fruitful discussions on these subjects. We would like to thank the support, knowledge, and encouragement from Carlile Lavour, Giuliano de la Guardia, Juliana Vizzotto, Marcelo Terra Cunha, Paulo Carreira Mateus, Renata Reiser, Renato Portugal, and Rubens Viana.

A very important person in this journey was our Springer editor Jorge Nakahara Jr. He took us by hand in this path to bring this book to life. His professional advices, patience, and continuous motivation were remarkable along the writing process. We are lucky to count on you!

We cannot forget some of our colleagues during our academic journey. We would like to thank Cheyenne Ribeiro, Francisco Revson Fernandes, Gilson Oliveira, and Edmar Nascimento.

The patience and support from our loved ones kept our hearts warm during the writing of this book. We are honored to have the quantum quintessence of your existence in our lives.

Contents

Acronyms

BEC	Binary Erasure Channel
BSC	Binary Symmetric Channel
CHSH	Clauser-Horne-Shimony-Holt
CSS	Calderbank-Shor-Steane
CZEC	Classical Zero-Error Capacity
DFS	Decoherence-Free Subspaces and Subsystems
DMC	Discrete Memoryless Channel
HSW	Holevo-Schumacher-Westmoreland
KS	Kochen-Specker
LOCC	Local Operations and Classical Communication
OSR	Operator-Sum Representation
POVM	Positive Operator-Valued Measurement
QEAC	Quantum Error-Avoiding Code
QECC	Quantum Error-Correcting Code
QMA	Quantum Merlin-Arthur
TPCP	Trace-Preserving Completely Positive
ZEAI	Zero-Error Accessible Information
ZESC	Zero-Error Secrecy Capacity

Chapter 1
Introduction

Classical information is, roughly speaking, everything that can be transmitted from a sender to a receiver with any finite set of symbols, like letters from the alphabet $(a, b, \ldots, z)$ or binary digits (0 or 1). When studying such kind of information we are not concerned with the semantic meaning of the information (such as in the complex and non-trivial meaning of the word "love"), but with syntactic and pragmatic aspects of it. The syntactic aspect deals with the relationship between the symbols used to construct a message and the pragmatic aspect that considers the actions taken by the parties (sender and receiver) in order to exchange messages [5]. This approach to deal with information was introduced by Claude Shannon in his seminal paper entitled "A Mathematical Theory of Communication" that inaugurated the *classical information theory* [17].

Classical information theory is a branch of applied Mathematics, Electrical Engineering, and Computer Science involving the quantification, storage, and communication of information. The results of this theory lie at the heart of every modern technology, underpinning all communications, networking, and data storage systems [4].

An interesting historical fact regarding this theory is that Claude Shannon both posted its fundamental problems and also, to a large extent, answered them. These problems were:

1. How much can a message be compressed?
2. At what rate can we communicate reliably over a noisy channel?

The answer to the first question is the *noiseless coding theorem*, whose objective is to remove redundancy, making messages smaller. It enabled many following results regarding data compression. The answer for the second question is the *channel coding theorem*, which gave rise to the entire fields of error-correcting codes and channel coding theory.

At the time, Shannon himself realized that his results would not only have applications in communication theory, but also in the theory of computing machines, the

© Springer International Publishing Switzerland 2016
E.B. Guedes et al., *Quantum Zero-Error Information Theory*,
DOI 10.1007/978-3-319-42794-2_1

design of telephone exchanges, and other fields [1]. In fact, besides communication theory, classical information theory results developed by Shannon and by many other researchers have fundamental contributions in the fields of Physics, Electrical Engineering, Computer Science, Probability and Statistics, among many others.

Until the 1970s, the realization of physical devices based on information and communication theory was restricted to apparatus whose operation was governed by the laws of classical physics. Such devices include telephones, radios, computers, and communication systems, among others. Especially from the 1980s, theoretical works together with experimental realizations have demonstrated that quantum mechanical features and properties could be used to build up communication systems. The most impressive example is the quantum cryptographical key distribution system described and realized by Bennett and Brassard in 1984 [2]. The quantum key distribution (QKD) protocol BB84, as the system became known, has caught the interest of many researchers around the world, giving rise to a new and exciting research area: the quantum information, communication, and computing theory. Although quantum mechanics theory started to be understood by the 1900s, as we see below, technological constraints had limited the development of devices based on this theory.

In 1900, Max Planck noticed that, upon studying black body radiation, particles like atoms and photons did not follow the rules of classical Physics. In order to overcome such problem, *quantum mechanics* was developed from 1900 to 1920, being considered nowadays as the most accurate model of reality that is currently known. It made possible the understanding of the fundamental particles and forces of nature, culminating in the development of the standard model of particle physics [14].

Since quantum mechanics describes the physics of quantum particles, what would change if we store information in a quantum system, such as a photon, or the spin of an electron? The answer to this question made *quantum information theory* come into existence. According to it, quantum information is information stored as a property of a quantum system that can be transmitted, stored, and processed following the laws of quantum mechanics.

Quantum information theory requires a basic understanding of quantum mechanics because quantum information has some unique characteristics, such as:

• The state of a quantum system cannot be copied or measured without disturbing it;
• The quantum state of two systems can be entangled, the two-system ensemble has a definite state, though neither individual system has a well-defined state of its own;
• It is not possible to reliably distinguish non-orthogonal states of a quantum system [5].

The properties of quantum information are remarkable and can be exploited for information processing. Quantum information theory may, thus, be defined as the study of the achievable limits to information processing possible within quantum mechanics [13]. We can also say that quantum information theory combines classical information theory with quantum mechanics to model information-related processes in quantum systems.

At a more fundamental level, it has become clear that an information theory based on quantum principles extends and complements classical information theory. The new theory includes quantum generalizations of classical notions such as sources, channels, and codes, as well as two complementary, quantifiable kinds of information—classical information and quantum entanglement [3].

Quantum information theory may be defined as the study of the achievable limits to information processing possible within quantum mechanics. First, it aims to determine limits on the class of information processing tasks which are possible in quantum mechanics. Second, it provides constructive means for achieving information processing tasks [13]. In a more detailed description, we can say that quantum information theory deals with four main topics [15]:

1. Transmission of classical information over quantum channels;
2. The tradeoff between acquisition of information about a quantum state and disturbance of the state;
3. Quantifying quantum entanglement;
4. Transmission of quantum information over quantum channels.

Regarding the first and the fourth topics which deal with information transmission, we are interested in understanding how classical information theory concepts and theorems are translated to the quantum domain in particular, because there are differences between classical and quantum channels. While classical channels can only transmit classical information, quantum channels can transmit classical information, private classical information, or quantum information. It can be used alone, with shared entanglement, or together with other classical and quantum channels.

It is possible to say that some results are well established in the quantum information theory literature. For example, quantum data compression, superdense coding, quantum teleportation, and entanglement concentration exemplify non-trivial ways in which quantum channels can be used, alone or in combination with classical channels, to transmit classical and quantum information. More recently, quantum error-correcting codes and entanglement distillation protocols have been discovered, which allow reliable transmission of classical and quantum information through noisy quantum channels [3].

One of the most intuitive ways to develop quantum information theory was considering a search, a translation, or an analogy of classical information theory concepts and theorems to the quantum domain. This approach allowed many successful results, such as in the case of the Schumacher's quantum noiseless coding theorem [16] which is a quantum counterpart of the Shannon's noiseless coding theorem.

Recalling the importance of Shannon's 1956 findings to the classical information theory, 8 years after his seminal paper, Shannon himself demonstrated how classical channels could be used to transmit information in a scenario where no errors are allowed [18]. This approach differs from his early results where an asymptotically small probability of error could be allowed. Considering this new scenario, he defined the so-called *zero-error capacity* which can be understood as the least upper bound of rates at which information can be transmitted through a classical channel with a probability of error equal to zero. In this domain where errors are not allowed, combinatorics and graph theory play an important role.

Considering an intuitive search for the equivalent of classical concepts in the quantum domain, Medeiros and de Assis proposed a quantum counterpart of the zero-error capacity, the *zero-error classical capacity of quantum channels* [7–12]. In their work, they established the conditions required to send classical information through quantum channels without decoding errors. With such a contribution, they inaugurated the field of *quantum zero-error information theory*.

Quantum zero-error information theory is a sub-field of quantum information theory which aims at studying and proposing techniques, protocols, and information measures to allow the transmission of classical or quantum information through noisy quantum channels without decoding errors.

After Medeiros' and de Assis' seminal work in quantum zero-error information theory, many contributions and results started to appear in the literature. Beigi and Shor studied the computational complexity of computing the zero-error capacity of quantum channels. They proposed the quantum clique problem and demonstrated that solving the later is equivalent to finding the zero-error capacity of a quantum channel. Beigi and Shor proved that the quantum clique problem is $\mathcal{QMA}$-complete. Developments from several authors showed that two quantum channels with a vanishing zero-error capacity can be combined in such ways that the joint channel has positive zero-error capacity, i.e., the zero-error capacity of quantum channels can be superactivated. This is an amazing result with no classical counterpart, since the superactivation is only possible, thanks to the quantum entanglement; the use of entangled states between two channel uses *activates* the capacity of two (originally incapable) quantum channels for transmitting information. Elloá and de Assis studied the secrecy capacity of quantum channels in a zero-error scenario, demonstrating that zero-error communication can be reached altogether with secrecy when communicating by means of quantum channels. Also, they defined a new measure of information for quantum sources, the so-called zero-error accessible information of a quantum source. Other remarkable results to this field are the generalization of the Lovász ϑ functional to quantum theory, the definition of new quantities in graph theory, and the study of their quantum computational complexity. This book presents these and other results on quantum zero-error information theory since its establishment in 2005.

We can see from the recent developments and results from many authors in the literature that we can indeed consider the quantum zero-error information as a theory. According to the Oxford dictionary, the meaning of the word "theory" can be understood as a set of principles on which the practice of an activity is based [6]. In our scenario, the principle considered is not to allow errors in information. All the results and developments observed so far are a consequence of this premise.

In the following pages, the reader will be guided from the basic concepts to the most recent findings in quantum zero-error information theory. We show how ideas that started with Shannon were updated and developed in order to consider quantum information in a scenario with complete absence of errors. The reader will be able to see how quantum zero-error information theory is interesting and worth exploring.

References

1. Aftab O, Cheung P, Kim A, Thakkar S, Yeddanapudi N (2001) Information theory and the digital age. Technical Report, Massachusetts Institute of Technology
2. Bennett CH, Brassard G (1984) Quantum cryptography: public key distribution and coin tossing. In: IEEE international conference on computers, systems and signal processing. IEEE Computer Society, Bangalore, pp 175–179
3. Bennett CH, Shor PW (1998) Quantum information theory. IEEE Trans Inform Theory 44(6):2724–2742
4. Desurvire E (2009) Classical and quantum information theory. Cambridge University Press, Cambridge
5. Marinescu DC, Marinescu GM (2011) Classical and quantum information, 1st edn. Academic, New York
6. McKean E (ed) (2010) The new Oxford American dictionary, 3rd edn. Oxford University Press, Oxford
7. Medeiros RAC (2008) Zero-error capacity of quantum channels. Ph.D Thesis, Universidade Federal de Campina Grande – TELECOM Paris Tech
8. Medeiros RAC, de Assis FM (2004) Zero-error capacity of a quantum channel. Lect Notes Comput Sci 3124:100–105
9. Medeiros RAC, de Assis FM (2005) Capacidade erro-zero de canais quânticos e estados puros. In: Simpósio Brasileiro de Telecomunicações, Campinas, São Paulo, pp 1–6
10. Medeiros RAC, de Assis FM (2005) Quantum zero-error capacity. Int J Quantum Inf 3(1):135–139
11. Medeiros RA, Alleaume R, Cohen G, de Assis FM (2006) Zero-error capacity of quantum channels and noiseless subsystems. In: IEEE International Telecommunications Symposium, Fortaleza, Brazil, pp 3–6
12. Medeiros RAC, Alleaume R, Cohen G, de Assis FM (2006) Quantum states characterization for the zero-error capacity. http://arxiv.org/abs/quant-ph/0611042. Accessed 25 Oct 2013
13. Nielsen MA (1998) Quantum information theory. Ph.D Thesis, University of New Mexico, Albuquerque, New Mexico, USA
14. Nielsen MA, Chuang IL (2010) Quantum computation and quantum information. Cambridge University Press, Cambridge
15. Preskill J (2015) Quantum information and computation. Lecture notes for Physics, vol. 229, 1st edn. CreateSpace Independent Publishing Platform, CA
16. Schumacher B (1995) Quantum coding. Phys Rev A 51:2738–2747
17. Shannon CE (1948) A mathematical theory of communication. Bell Syst Tech J 27:623–656
18. Shannon CE (1956) The zero error capacity of a noisy channel. IRE Trans Inf Theory 2(3):8–19

Chapter 2
Fundamentals of Quantum Information Processing

Quantum mechanics is a part of quantum theory that aims at describing the nature when we consider the subatomic physics, or *quantum physics*. It can also be understood as the mathematical framework to describe isolated quantum systems, whose behavior cannot be captured by *classical physics* [10]. We have two ways to consider when we have the necessity to incorporate quantum mechanical effects into computing and communications: (1) use it to strive to suppress the quantum effects and still preserve a semblance of classicality even though the computational or communication elements are very small; or (2) use it to enhance quantum effects and try to find clever ways to enhance and sustain them to achieve old computational and communication goals in new ways. *Quantum computing* and *communications* use the latest strategy by harnessing quintessentially quantum effects [13].

Taking into account the approach required for quantum computation and communications, when we desire to build algorithms and hardware for quantum computing and communications we must consider the *postulates of quantum mechanics*. These postulates specify how we can represent, process, and measure information in this new domain.

In this chapter we are going to introduce some basic concepts of quantum information processing. Firstly, we make the reader familiar with the Dirac notation [1], a concise representation of the quantum mechanics concepts, which implies in a simplification of the calculi to be performed. After that, we are going to introduce the *density operators*, very useful in the domain of quantum communications.

This chapter provides a quick review of basic concepts for the understanding of subsequent chapters. However, it does not contain a complete explanation of the mathematics behind quantum mechanics. It is extensive and there are entire books dedicated to it. The chapter concludes with suggestions for further reading.

This chapter is organized as follows. Section 2.1 introduces the qubit, which is the fundamental unit of information in a quantum system. Section 2.2 describes how the evolution is carried out in a quantum system, whereas Sect. 2.3 shows how we can bring information from a quantum system to a classical level by means of

© Springer International Publishing Switzerland 2016
E.B. Guedes et al., *Quantum Zero-Error Information Theory*,
DOI 10.1007/978-3-319-42794-2_2

projective measurements and of positive operator-value measurements. The density operator formalism, very useful for quantum communications, is shown in Sect. 2.4. Entanglement, which does not have a classical counterpart, is discussed in Sect. 2.5. The postulates of quantum mechanics are presented in Sect. 2.6 using the density operator notation.

2.1   Representing Information

The basic unity of information in classical computing and communications is the *bit*, or *binary digit*, which can be 0 or 1 (false or true, respectively). The information is represented, or encoded, by a sequence of bits.

On the other hand, the basic unity of information in quantum computing and communications is a two-state quantum mechanical system: the *qubit*, or *quantum bit*. Consequently, the state of a qubit is represented by a unit vector in a 2-dimensional complex Hilbert space. We call such a vector a *ket* and we denote the state by $|\psi\rangle$, where $|\cdot\rangle$ is the vector and ψ is the label of the qubit. The following definition formalizes the state of a qubit.

Definition 2.1 (State of a Qubit). The state of a qubit ψ, denoted by $|\psi\rangle$, can be represented by a vector in a 2-dimensional Hilbert space $\mathcal{H}$, i.e.,

$$|\psi\rangle = \alpha\,|0\rangle + \beta\,|1\rangle, \tag{2.1}$$

where α and β are complex numbers ($\alpha, \beta \in \mathbb{C}$), which must satisfy the unitary restriction

$$|\alpha|^2 + |\beta|^2 = 1. \tag{2.2}$$

The states $|0\rangle$ and $|1\rangle$ form a basis for the 2-dimensional Hilbert space.

The use of $|\psi\rangle$ to represent the state of a qubit does not depend on the manner that it is physically encoded. A certain state $|\phi\rangle$, for example, could refer to the state of a polarized photon, or an excited state of an atom, or the direction of circulation of a superconducting current, etc. [13]. Such representation enables us to treat qubits as abstract mathematical objects [10, Chap. 1].

In (2.1), the states $|0\rangle$ and $|1\rangle$ are known as *computational basis states*, and form an orthonormal basis for the 2-dimensional Hilbert space. The notation of $|0\rangle$ and $|1\rangle$ in terms of vectors is

$$|0\rangle = \begin{bmatrix} 1 \\ 0 \end{bmatrix}, \qquad |1\rangle = \begin{bmatrix} 0 \\ 1 \end{bmatrix}. \tag{2.3}$$

Using the computational basis, the general state of a qubit $|\psi\rangle$ is denoted by

$$|\psi\rangle = \alpha |0\rangle + \beta |1\rangle$$

$$= \alpha \begin{bmatrix} 1 \\ 0 \end{bmatrix} + \beta \begin{bmatrix} 0 \\ 1 \end{bmatrix}$$

$$= \begin{bmatrix} \alpha \\ \beta \end{bmatrix}. \tag{2.4}$$

We say that α and β are the *amplitudes* associated with the states $|0\rangle$ and $|1\rangle$, respectively.

When α and β in (2.1) are non-zero, we say that the qubit is in a *superposition* state. Differently from bits, qubits are not constrained to be wholly 0 or wholly 1 at a given moment [13]. According to quantum mechanics, the modulus squared of α, β in the former equation gives the probability of measuring the qubit in state $|0\rangle$ or in state $|1\rangle$, respectively [7]. It means that

- $|\alpha|^2$ gives the probability of finding $|\psi\rangle$ in state $|0\rangle$;
- $|\beta|^2$ gives the probability of finding $|\psi\rangle$ in state $|1\rangle$.

The unitary restriction of Definition 2.1 is related to the probability of obtaining a given measurement result. In particular, if $|\alpha| = |\beta|$, we say that the qubit is in an equally distributed superposition.

Example 2.1 (Qubits in Superposition). Let $|\varphi\rangle$ and $|\phi\rangle$ denote the states of two qubits in superposition:

$$|\varphi\rangle = \frac{1}{\sqrt{3}} |0\rangle + \sqrt{\frac{2}{3}} |1\rangle, \tag{2.5}$$

$$|\phi\rangle = \frac{1}{\sqrt{2}} |0\rangle - \frac{1}{\sqrt{2}} |1\rangle. \tag{2.6}$$

Although both qubits are in superposition, only the state $|\phi\rangle$ is in an equally distributed superposition.

The Hermitian conjugate of a ket $|\psi\rangle$ is called a *bra* or *dual vector*. A bra is denoted by $\langle\psi| = |\psi\rangle^\dagger$, where † indicates complex conjugation and matrix transposition.

$$\langle\psi| = |\psi\rangle^\dagger$$

$$= \left(|\psi\rangle^*\right)^T$$

$$= \begin{bmatrix} \alpha^* \\ \beta^* \end{bmatrix}^T$$

$$= \begin{bmatrix} \alpha^* & \beta^* \end{bmatrix}. \tag{2.7}$$

Example 2.2 (Bra). The Hermitian conjugate of (2.5) is given by

$$\langle\varphi| = |\varphi\rangle^{\dagger}$$

$$= \begin{bmatrix} \frac{1}{\sqrt{3}} \\ \sqrt{\frac{2}{3}} \end{bmatrix}^{\dagger}$$

$$= \begin{bmatrix} \frac{1}{\sqrt{3}}^* & \sqrt{\frac{2}{3}}^* \end{bmatrix}$$

$$= \begin{bmatrix} \frac{1}{\sqrt{3}} & \sqrt{\frac{2}{3}} \end{bmatrix}.$$

A remarkable difference from quantum to classical computing and communication is that while a bit can represent only two distinct values, a qubit can assume infinitely many different states. The only constraint is the unitary restriction (2.1). Therefore, while the basic unity of quantum information is unlimited, the classical unity of information is restricted to the values "true" and "false."

We can also introduce a geometrical representation for a single qubit. To do so, we rewrite (2.1) as

$$|\psi\rangle = e^{i\gamma}\left[\cos\left(\frac{\alpha}{2}\right)|0\rangle + e^{i\beta}\sin\left(\frac{\alpha}{2}\right)|1\rangle\right], \tag{2.8}$$

where $\alpha, \beta, \gamma \in \mathbb{R}$. Factor $e^{i\gamma}$ is called the global phase. The factor does not influence measurement statistics, since its absolute value is equal to one. Consequently, global phase is often omitted [5]. While (2.1) is related to a vector in a two-dimensional Hilbert space, the qubit (2.8) without $e^{i\gamma}$ has a nice geometrical interpretation on a three-dimensional polar coordinate system. Figure 2.1 illustrates the representation of (2.8) in a *Bloch sphere*.

Example 2.3 (Geometrical Representation). The qubit (2.6) can be denoted according to the geometrical representation as

$$|\phi\rangle = \cos\left(\frac{\pi}{4}\right)|0\rangle - \sin\left(\frac{\pi}{4}\right)|1\rangle.$$

2.1.1 Composite Quantum Systems

In classical computing and communications, a bit can represent two values, 0 or 1. Therefore, a register of n bits can store 2^n different values, one each time. Thanks to the superposition of quantum states, a quantum register of n qubits can store 2^n different values at the same time. The concept of composite quantum systems is shown in Definition 2.2.

Fig. 2.1 Geometrical
visualization of a qubit in
Bloch sphere

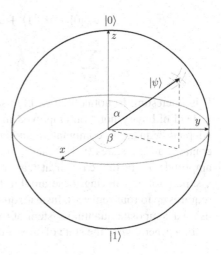

Definition 2.2 (Composite Quantum Systems). A *composite quantum system*,
also called quantum register or multi-qubit quantum systems, is made up of two
or more distinct physical systems. The state space of a composite quantum system
is the tensor product of the state space of its components. If $|\psi_1\rangle, \ldots, |\psi_n\rangle$ describe
the states of n isolated quantum systems, the state of the composite system is
$|\psi_1\rangle \otimes \ldots \otimes |\psi_n\rangle$.

The tensor product of the states $|a\rangle$ and $|b\rangle$, also known as direct or Kronecker
product, is denoted by $|a\rangle \otimes |b\rangle = |ab\rangle$, and calculated as follows:

$$
|a\rangle \otimes |b\rangle = \begin{bmatrix} a_1 \\ a_2 \\ \cdots \\ a_n \end{bmatrix} \otimes \begin{bmatrix} b_1 \\ b_2 \\ \cdots \\ b_n \end{bmatrix}
$$

$$
= \begin{bmatrix} a_1 \cdot |b\rangle \\ a_2 \cdot |b\rangle \\ \cdots \\ a_n \cdot |b\rangle \end{bmatrix}. \tag{2.9}
$$

Let $|\psi\rangle$ be the state of a certain 2-qubit system. The representation of $|\psi\rangle$ is
given by

$$
\begin{aligned}
|\psi\rangle &= |\psi_1\rangle \otimes |\psi_2\rangle \\
&= (\alpha_1 |0\rangle + \beta_1 |1\rangle) \otimes (\alpha_2 |0\rangle + \beta_2 |1\rangle) \\
&= \alpha_1\alpha_2 |00\rangle + \alpha_1\beta_2 |01\rangle + \beta_1\alpha_2 |10\rangle + \beta_2\beta_2 |11\rangle \\
&= \alpha_{00} |00\rangle + \alpha_{01} |01\rangle + \alpha_{10} |10\rangle + \alpha_{11} |11\rangle
\end{aligned} \tag{2.10}
$$

$$= \alpha'_0 |0\rangle + \alpha'_1 |1\rangle + \alpha'_2 |2\rangle + \alpha'_3 |3\rangle$$

$$= \sum_{i=0}^{2^2-1} \alpha'_i |i\rangle . \tag{2.11}$$

It is interesting to notice that (2.11) uses decimal notation for indexes and labels instead of binary notation employed in (2.10)—this is a simplification commonly adopted in quantum computation and communication. The state $|\psi\rangle$ of two qubits contains all the states 0, 1, 2, and 3 at the same time, each of them with its own amplitude α'_i. In this case, if at least two distinct $\alpha'_i \neq 0$, we say that $|\psi\rangle$ is in a superposition. Storing these amplitudes on a classical computer simultaneously requires up to four registers. Instead, quantum computers perform the same task just using a composite quantum system of two qubits in superposition.

In a general way, the state of an n-qubit system can be written as

$$|\psi\rangle = \sum_{i=0}^{2^n-1} \alpha_i |i\rangle , \tag{2.12}$$

with $\sum_{i=0}^{2^n-1} |\alpha_i|^2 = 1$. The states $|0\rangle, |1\rangle, \ldots, |2^n - 1\rangle$ form the computational basis of the 2^n-dimensional Hilbert space.

Another way to represent a composite quantum state is to consider it as belonging to a larger Hilbert space $\mathcal{H}$ composed by tensor product of state vectors belonging to Hilbert spaces $\mathcal{H}_1$ and $\mathcal{H}_2$. We can construct a certain vector $|\psi\rangle \in \mathcal{H}$ as the tensor product of two vectors $|\phi\rangle \in \mathcal{H}_1$ and $|\varphi\rangle \in \mathcal{H}_2$:

$$|\psi\rangle = |\phi\rangle \otimes |\varphi\rangle . \tag{2.13}$$

The tensor product has the following properties:

1. For $\alpha \in \mathbb{C}$, $|\phi\rangle \in \mathcal{H}_1$ and $|\psi\rangle \in \mathcal{H}_2$,

$$\alpha \left(|\phi\rangle \otimes |\psi\rangle \right) = \left(\alpha |\phi\rangle \right) \otimes |\psi\rangle = |\phi\rangle \otimes \left(\alpha |\varphi\rangle \right) . \tag{2.14}$$

2. For $|\phi_1\rangle, |\phi_2\rangle \in \mathcal{H}_1$ and $|\psi\rangle \in \mathcal{H}_2$,

$$\left(|\phi_1\rangle + |\phi_2\rangle \right) \otimes |\psi\rangle = |\phi_1\rangle \otimes |\psi\rangle + |\phi_2\rangle \otimes |\psi\rangle . \tag{2.15}$$

3. For $|\phi\rangle \in \mathcal{H}_1$ and $|\psi_1\rangle, |\psi_2\rangle \in \mathcal{H}_2$,

$$|\phi\rangle \otimes \left(|\psi_1\rangle + |\psi_2\rangle \right) = |\phi\rangle \otimes |\psi_1\rangle + |\phi\rangle \otimes |\psi_2\rangle . \tag{2.16}$$

2.2 Processing Information

Classical information processing is performed by applying operations on bits that represent information. In the quantum scenario, the information processing is also performed by *operators*, which are applied on the state of qubits. Typically, these operators are denoted by capital letters of the alphabet and have special properties stated in the following definition:

Definition 2.3 (Quantum Operator). An isolated quantum system originally in the state $|\psi_1\rangle$ *evolves* to state $|\psi_2\rangle$ by means of the application of a *quantum operator U*:

$$|\psi_2\rangle = U |\psi_1\rangle. \tag{2.17}$$

Quantum operators are required to be *unitary* because they should preserve vector norms. A unitary operator U has the following property: $U^\dagger = U^{-1}$, where $U^\dagger$ denotes the Hermitian conjugate (conjugate transpose) of U and U^{-1} is the inverse of U. Therefore, any unitary operator satisfies

$$U^\dagger \cdot U = U \cdot U^\dagger = \mathbb{1}, \tag{2.18}$$

where $\mathbb{1}$ is the identity matrix [6].

Because quantum operators are unitary, the evolution of an isolated quantum system is *reversible*. For example, we can easily return to the state $|\psi_1\rangle$ from $|\psi_2\rangle$ just applying the unitary operator $U^\dagger$:

$$U^\dagger |\psi_2\rangle = U^\dagger (U |\psi_1\rangle) = (U^\dagger U) |\psi_1\rangle = |\psi_1\rangle. \tag{2.19}$$

Some operators play an important role in quantum information processing and quantum computing. Particularly, the set known as *Pauli matrices* is specially interesting:

$$\sigma_0 = \mathbb{1} = \begin{bmatrix} 1 & 0 \\ 0 & 1 \end{bmatrix}, \qquad\qquad \sigma_1 = \sigma_x = X = \begin{bmatrix} 0 & 1 \\ 1 & 0 \end{bmatrix},$$

$$\sigma_2 = \sigma_y = Y = \begin{bmatrix} 0 & -i \\ i & 0 \end{bmatrix}, \qquad\qquad \sigma_3 = \sigma_z = Z = \begin{bmatrix} 1 & 0 \\ 0 & -1 \end{bmatrix}.$$

The Pauli matrices are Hermitian, i.e., $\sigma_k = \sigma_k^\dagger$, $k = 0, \ldots, 3$. The operator X, in particular, is the quantum analog of the classical *NOT* gate. For instance, when this operator is applied to $|0\rangle$ we get $X |0\rangle = |1\rangle$. Pauli matrices are widely used in several quantum computation and communication algorithms.

The *Hadamard matrix*, denoted by H, is another very important operator because it can build equally distributed superpositions. Moreover, when applied to any state of a 2-dimensional Hilbert space, it performs a $45°$ rotation on such state. The matricial representation of this operator is

$$H = \frac{1}{\sqrt{2}} \begin{bmatrix} 1 & 1 \\ 1 & -1 \end{bmatrix}. \tag{2.20}$$

If we apply Hadamard to the state $|1\rangle$, this operator will create the superposition $H|1\rangle = \frac{1}{\sqrt{2}}(|0\rangle - |1\rangle) = |-\rangle$. The states $|-\rangle$ and $|+\rangle = H|0\rangle$ are known as the *Hadamard basis*.

2.2.1 Tensor Product of Operators

Suppose we have a composite quantum system and we wish to apply a quantum operator to each of the respective states. To enable quantum operators to be applied in multi-qubit systems, we must define tensor products of quantum operators.

Definition 2.4 (Tensor Product of Operators). Let A and B be the matricial representation of two quantum operators with dimensions $m \times n$ and $p \times q$, respectively. The tensor product $A \otimes B$ is defined by

$$A \otimes B \equiv \begin{bmatrix} A_{11}B & A_{12}B & \ldots & A_{1n}B \\ A_{21}B & A_{22}B & \ldots & A_{2n}B \\ \vdots & \vdots & \ddots & \vdots \\ A_{m1}B & A_{m2}B & \ldots & A_{mn}B \end{bmatrix}. \tag{2.21}$$

The resulting matrix $A \otimes B$ has dimension $(nq) \times (mp)$ and can be applied to a composite quantum system as previously explained. We denote the n-tensor product of the operator U with itself by $U^{\otimes n}$. For example, $U^{\otimes 3} = U \otimes U \otimes U$.

The Hadamard operator, in particular, is very useful in many quantum computing and communication algorithms. For example, n-tensor product of Hadamard operators can be used to create n equally distributed superposition of qubits. Then, an arbitrary quantum operator acting on the state space of the n-qubits can be applied to the composite system in a simultaneous way. This feature, called *quantum parallelism*, is restricted to quantum computation. Parallelism is not performed efficiently by classical computers because, for instance, simulating a superposition of n qubits requires 2^n classical registers and individual application of the operation in each of them.

Example 2.4 (Tensor Product of Operators). Let $|\psi\rangle = |00\rangle = |0\rangle^{\otimes 2}$ be a 2-qubit quantum system. The application of the Hadamard operator to both qubits is performed by the operator $H^{\otimes 2} = H \otimes H$ in the following way:

$$H^{\otimes 2} |0\rangle^{\otimes 2} = H |0\rangle \otimes H |0\rangle$$

$$= \left[\frac{1}{\sqrt{2}} (|0\rangle + |1\rangle) \right] \otimes \left[\frac{1}{\sqrt{2}} (|0\rangle + |1\rangle) \right]$$

$$= |+\rangle |+\rangle$$

$$= |+\rangle^{\otimes 2}.$$

2.2.2 Projection Operators

The result of the outer product operation on a vector $|\psi\rangle$ with itself, denoted by $|\psi\rangle \langle \psi|$, is a linear projection operator. Such operator $|\psi\rangle \langle \psi|$, performs the following mapping:

$$|\psi\rangle \langle \psi | |\varphi\rangle \mapsto |\psi\rangle \langle \psi | \varphi\rangle = \langle \psi | \varphi\rangle |\psi\rangle , \qquad (2.22)$$

where $|\psi\rangle , |\varphi\rangle \in \mathcal{H}$. That is, the operator $|\psi\rangle \langle \psi|$ projects a vector $|\varphi\rangle$ onto the 1-dimensional subspace of $\mathcal{H}$ spanned by $|\psi\rangle$. Such an operator is called an *orthogonal projector* [6].

More generally, suppose $\mathcal{H}$ is an n-dimensional Hilbert space. Let $\{|1\rangle , \ldots , |k\rangle\}$ be any orthonormal basis of a subspace $\mathcal{H}'$ of $\mathcal{H}$, $k \leq n$. Then,

$$P = \sum_{i=1}^{k} |i\rangle \langle i| \qquad (2.23)$$

is an orthogonal projector onto the subspace $\mathcal{H}'$. It is easy to see that projectors are Hermitian operators. Moreover, for any orthogonal projector P, $P^2 = P$ [7].

Projection operators also satisfy the *completeness relation*, i.e., if $\{|1\rangle , \ldots , |n\rangle\}$ is an orthonormal basis of an n-dimensional Hilbert space $\mathcal{H}$, then

$$P = \sum_{i=1}^{n} |i\rangle \langle i| = \mathbb{1}. \qquad (2.24)$$

Example 2.5 (Completeness Relation). Let $B_H = \{|+\rangle , |-\rangle\}$ be the Hadamard basis of the 2-dimensional Hilbert space. The corresponding projection operators are

$$P_+ = |+\rangle \langle +| = \begin{bmatrix} \frac{1}{2} & \frac{1}{2} \\ \frac{1}{2} & \frac{1}{2} \end{bmatrix},$$

$$P_- = |-\rangle \langle -| = \begin{bmatrix} \frac{1}{2} & -\frac{1}{2} \\ -\frac{1}{2} & \frac{1}{2} \end{bmatrix}.$$

It is straightforward to see that $P_+ + P_- = \mathbb{1}$.

2.3 Measuring Information

An isolated quantum system evolves by means of unitary transformations. While it remains closed, no information can be inferred from the system. In order to access the state of a quantum system, we need to perform a task called *measurement*. Measurements can be viewed as an "interface" from the quantum world to the classical level; it is the unique way to extract useful information from qubits after some processing.

In the classical scenario, measurement is a trivial task and the results depend only on the apparatus accuracy. Moreover, measurements do not disturb the state of the classical system, no matter how many times we measure the corresponding quantity. However, in quantum scenario, measurement is not a trivial task because it affects the isolated quantum system causing a collapse in the state space of corresponding quantum system being measured. As a consequence, measurements are irreversible operations in quantum systems—once a qubit is measured, it is not possible to return to the state it had right before the measurement [8].

A classical computer follows essentially a load-run-read cycle wherein one loads data into the machine, runs a program using this data as input, and then reads out the result. This becomes an analogous prepare-evolve-measure cycle on a quantum computer. That is, one prepares a quantum state, evolves it on the quantum computer by means of unitary transformations and, finally, measures the result [13].

There are two special cases of general measurements that play an important role in quantum information and computation: projective measurements and positive operator-valued measurements.

Definition 2.5 (Projective Measurement). A projective measurement is described by an observable M, which is a Hermitian operator on the state space of the system being measured. The observable M has a spectral decomposition

$$M = \sum_m \lambda_m P_m, \tag{2.25}$$

where P_m is a projector onto the eigenspace of M with eigenvalue λ_m. Measurement outcomes correspond to the eigenvalue indexes m. When a system in a state $|\psi\rangle$ is observed, the probability of getting output m is

$$p(m) = \langle \psi | P_m | \psi \rangle. \tag{2.26}$$

Given that the outcome m occurred, the state of the system immediately after the measurement will be

$$|\psi'\rangle = \frac{P_m |\psi\rangle}{\sqrt{p(m)}}. \tag{2.27}$$

Instead of giving an observable to describe a projective measurement, we can simply construct a list of projectors P_m satisfying $\sum_m P_m = \mathbb{1}$ and $P_i P_j = \delta_{ij} P_i$, i.e., projectors must be pairwise orthogonal. The corresponding observable is then $M = \sum_m m P_m$. We say that a quantum system is measured in a basis $|m\rangle$ when a projective measurement with projectors $P = |m\rangle\langle m|$ is performed, where $|m\rangle$ is an orthonormal basis.

Example 2.6 (Projective Measurements). Suppose a quantum system in the state $|\psi\rangle = \frac{1}{\sqrt{2}}(|0\rangle - |1\rangle)$. Performing a projective measurement in the computational basis with projectors $\{P_0 = |0\rangle\langle 0|, P_1 = |1\rangle\langle 1|\}$ gives the output "0" with probability

$$
\begin{aligned}
p(0) &= \langle\psi| P_0 |\psi\rangle \\
&= \left(\frac{\langle 0| - \langle 1|}{\sqrt{2}}\right) |0\rangle\langle 0| \left(\frac{|0\rangle - |1\rangle}{\sqrt{2}}\right) \\
&= \frac{1}{2}.
\end{aligned}
$$

Similarly, we found that $p(1) = \frac{1}{2}$. In this case, getting the two possible outputs is an equally likely event. Given that outcome "0" occurs, the post measurement state will be

$$
\begin{aligned}
|\psi'\rangle &= \frac{P_0 |\psi\rangle}{\sqrt{p(0)}} \\
&= \frac{|0\rangle\langle 0| \left(\frac{|0\rangle - |1\rangle}{\sqrt{2}}\right)}{\sqrt{\frac{1}{2}}} \\
&= |0\rangle.
\end{aligned}
$$

As we can expect, the post-measurement state given that output "1" occurred is $|\psi'\rangle = |1\rangle$.

In some applications, however, the system state after the measurement is not important. For example, in quantum error-correction codes, the measurement output on the received quantum codeword gives the error syndrome, which is used to choose a unitary operator in order to (possibly) correct the error introduced by the noisy quantum channel. In such situations, we are only interested in the outcomes and their associated probabilities. The *Positive Operator-Value Measurement formalism* (POVM formalism) is the most appropriate theoretical tool to deal with this scenario.

Definition 2.6 (POVM Measurements). A Positive Operator-Value Measurement (POVM) is defined by a set of Hermitian, positive operators $\{E_m\}$ acting on the state space of the quantum system being measured [10, Sect. 2.2.6]. The probability of

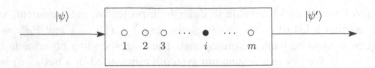

Fig. 2.2 A POVM measurement apparatus. When a quantum state is measured using a set of POVM elements $\{E_1, \ldots, E_m\}$, an lcd is turned on indicating the outcome

getting outcome m given that the state $|\psi\rangle$ is measured is

$$p(m) = \langle\psi| E_m |\psi\rangle. \tag{2.28}$$

POVM operators must satisfy the completeness relation, i.e.,

$$\sum_m E_m = \mathbb{1}. \tag{2.29}$$

Differently from general and projective measurements, we are not able to predict the post-measurement state of quantum system after a POVM measurement. Fortunately, most of the applications in quantum computation and information do not care about post-measurement states. Instead, we are often interested in measurement outcomes and the corresponding associated probabilities. Figure 2.2 illustrates a POVM measurement apparatus. When an unknown quantum state $|\psi\rangle$ is measured, a led turns on to indicate the outcome.

2.4   Density Operator

Pure quantum states are represented by unitary vectors belonging to an appropriate Hilbert space. This kind of system suggests a lowest degree of ignorance, since we have nothing further to discover than the quantum state itself.

However, a qubit can be in an *ensemble* of pure states, i.e., the system can be in a certain state $|\psi_i\rangle$ with probability p_i, $i > 1$. We describe the state of such qubit as an ensemble of possible pure states and their associated probabilities $\{|\psi_i\rangle, p_i\}$, where $\sum_i p_i = 1$. The whole system is said to be in a *mixed quantum state*. In summary, the formalism we have used so far is not adequate to represent quantum systems in two situations:

1. When the quantum system state is one of $|\psi_1\rangle$, $|\psi_2\rangle$, $\ldots$ with probabilities p_1, p_2, $\ldots$.
2. When a certain system (called A) is part of a larger quantum system AB.

In these situations, the mathematical formalism of *density operators* is more suitable to describe the state of the whole quantum system.

Definition 2.7 (Density Operator). Suppose that a quantum system is in one of the states $|\psi_i\rangle$ with probability p_i, $\sum_i p_i = 1$. We say that the quantum system is an ensemble $\{|\psi_i\rangle, p_i\}$. The density operator that describes the whole system is defined as

$$\rho = \sum_i p_i |\psi_i\rangle \langle\psi_i|. \tag{2.30}$$

Example 2.7 (Density Operator—Pure State). Suppose that we apply the Hadamard operator to a quantum state $|\psi_0\rangle$ initially on the state $|0\rangle$. Then, the state of the quantum system will be

$$|\psi_1\rangle = H |\psi_0\rangle$$

$$= H |0\rangle$$

$$= \frac{1}{\sqrt{2}} (|0\rangle \langle 0| + |0\rangle \langle 1| + |1\rangle \langle 0| - |1\rangle \langle 1|) |0\rangle$$

$$= \frac{1}{\sqrt{2}} (|0\rangle + |1\rangle).$$

Using the density operator formalism, the system state after the Hadamard operation can be denoted as

$$\rho = |\psi_1\rangle \langle\psi_1|$$

$$= \left[\frac{1}{\sqrt{2}} (|0\rangle + |1\rangle)\right] \left[\frac{1}{\sqrt{2}} (\langle 0| + \langle 1|)\right]$$

$$= \frac{1}{2} (|0\rangle \langle 0| + |0\rangle \langle 1| + |1\rangle \langle 0| + |1\rangle \langle 1|)$$

$$= \frac{1}{2} \begin{bmatrix} 1 & 1 \\ 1 & 1 \end{bmatrix}.$$

Although we had used the density operator formalism to represent the state of the system, the system itself remains in a quantum pure state.

Example 2.8 (Density Operator—Mixed State). Consider that a quantum system can be in one of the states $|+\rangle$ and $|-\rangle$ with probability 1/3 and 2/3, respectively. The density operator of the system is given by

$$\rho = \left[\frac{1}{3} \left(\frac{|0\rangle + |1\rangle}{\sqrt{2}}\right) \left(\frac{\langle 0| + \langle 1|}{\sqrt{2}}\right)\right] + \left[\frac{2}{3} \left(\frac{|0\rangle - |1\rangle}{\sqrt{2}}\right) \left(\frac{\langle 0| - \langle 1|}{\sqrt{2}}\right)\right]$$

$$= \frac{1}{6} (|0\rangle \langle 0| + |0\rangle \langle 1| + |1\rangle \langle 0| + |1\rangle \langle 1|) +$$

$$+ \frac{2}{6} \left(|0\rangle \langle 0| - |0\rangle \langle 1| - |1\rangle \langle 0| + |1\rangle \langle 1| \right)$$

$$= \frac{|0\rangle \langle 0|}{2} - \frac{|0\rangle \langle 1|}{3} - \frac{|1\rangle \langle 0|}{3} + \frac{|1\rangle \langle 1|}{2}$$

$$= \begin{bmatrix} 1/2 & -1/3 \\ -1/3 & 1/2 \end{bmatrix}.$$

Density operators have a well-defined characterization. The reader can easily prove the following theorem:

Theorem 2.1 (Density Operator). *An operator ρ is a density operator associated with an ensemble $\{|\psi_i\rangle, p_i\}$ if and only if it satisfies two conditions:*

1. *Trace Condition. ρ has trace equal to 1;*
2. *Positivity Condition. ρ is a positive operator.*

Example 2.9 (Trace and Positivity). From the density operator ρ of the previous example, we can see that its trace is equal to 1, as stated by the trace condition.

$$\mathrm{Tr}(\rho) = \mathrm{Tr} \left(\begin{bmatrix} 1/2 & -1/3 \\ -1/3 & 1/2 \end{bmatrix} \right)$$

$$= \frac{1}{2} + \frac{1}{2}$$

$$= 1.$$

Positivity condition can be checked by calculating the eigenvalues of ρ, which are $\lambda_1 = \frac{5}{6}$ and $\lambda_2 = \frac{1}{6}$. Since both eigenvalues are positive, ρ is a positive operator as well.

Given a density matrix ρ, how can we infer that the corresponding quantum system is in a pure or mixed state? It turns out that all we need to do is calculate the trace of ρ^2, as shown in the following theorem.

Theorem 2.2 (Condition to ρ Describe a Pure or Mixed State). *Let ρ be a density operator representing a quantum system. Then, $\mathrm{Tr}\left(\rho^2\right) \leq 1$, with equality if and only if the system is in a pure state.*

The two previous examples are useful to illustrate the theorem. Density operator of Example 2.7 represents a quantum system in a pure state. Since

$$\rho = \frac{1}{2} \begin{bmatrix} 1 & 1 \\ 1 & 1 \end{bmatrix},$$

then

$$\text{Tr}\left(\rho^2\right) = \text{Tr}\left(\frac{1}{4}\begin{bmatrix} 2 & 2 \\ 2 & 2 \end{bmatrix}\right)$$

$$= 1.$$

Similarly, for the density operator of Example 2.8,

$$\text{Tr}\left(\rho^2\right) = \text{Tr}\left(\begin{bmatrix} 13/36 & -1/3 \\ -1/3 & 13/36 \end{bmatrix}\right)$$

$$= 13/18$$

$$< 1,$$

which means that the quantum system is in a mixed state.

2.5   Entanglement

Quantum systems display properties that are unknown for classical ones, such as the superposition of quantum states, interference, or tunneling. These are all one-particle effects that can be observed in quantum systems, which are composed of a single particle. But these are not the only distinctions between classical and quantum objects—there are further differences that manifest themselves in composite quantum systems, that is, systems that are comprised of at least two subsystems [9].

Entanglement is a property of two or more quantum systems which exhibit correlations that cannot be explained by classical physics [12], being a key resource in quantum computation and quantum information theory. Entanglement occurs on composite quantum systems and involves unusually strong correlation between parts of them [13]. We begin by defining an entangled pure state.

Definition 2.8 (Entangled Pure State). A multi-qubit pure state is entangled if and only if it cannot be factored into the direct product of a definite state for each qubit individually. Thus, a pair of qubits, A and B, are entangled if and only if their joint state $|\psi_{AB}\rangle$ cannot be written as the product of a pure state for qubit A and a pure state for qubit B, i.e., $|\psi_{AB}\rangle \neq |\psi_A\rangle \otimes |\psi_B\rangle$ for any choice of states $|\psi_A\rangle$ and $|\psi_B\rangle$ [13].

If the systems A and B are entangled, this means that the values of certain properties of system A are correlated with the values that those properties will assume for system B. The properties can become correlated even when the two systems are spatially separated [7].

Example 2.10 (Entangled Pure State [12]). States of two quantum systems can be considered together by taking their tensor product. For example, the two Hadamard states $|+\rangle$ and $|-\rangle$ can be considered together in the form

$$|+\rangle \otimes |-\rangle = \left[\frac{1}{\sqrt{2}} (|0\rangle + |1\rangle)\right] \otimes \left[\frac{1}{\sqrt{2}} (|0\rangle - |1\rangle)\right]$$

$$= \frac{1}{2} (|00\rangle + |01\rangle + |10\rangle - |11\rangle).$$

The state $|+\rangle \otimes |-\rangle$ represents a 2-qubit system which is not entangled. On the other hand, consider the following quantum state:

$$|\beta_{00}\rangle = \frac{|00\rangle + |11\rangle}{\sqrt{2}}. \tag{2.31}$$

If we try to write $|\beta_{00}\rangle$ as a tensor product of two pure qubits $|\psi_A\rangle = \alpha |0\rangle + \beta |1\rangle$ and $|\psi_B\rangle = \gamma |0\rangle + \delta |1\rangle$, we get

$$|\beta_{00}\rangle = (\alpha |0\rangle + \beta |1\rangle) \otimes (\gamma |0\rangle + \delta |1\rangle)$$

$$= \alpha\gamma |00\rangle + \beta\gamma |01\rangle + \alpha\delta |01\rangle + \beta\delta |11\rangle.$$

It is easy to see that we cannot find α, β, γ, and δ that simultaneously satisfy $\alpha\gamma = \beta\delta = 1$ and $\beta\gamma = \alpha\delta = 0$. Therefore, the state $|\beta_{00}\rangle$ is entangled, since it cannot be broken down into two separate qubit pure states.

The state $|\beta_{00}\rangle$ belongs to a very important set of entangled states known as *Bell states* or *EPR pairs*. The Bell states form a basis for the 4-dimensional Hilbert space and play an important role in many quantum communication protocols. The Bell states are defined as

$$|\beta_{00}\rangle = \frac{|00\rangle + |11\rangle}{\sqrt{2}}, \tag{2.32}$$

$$|\beta_{01}\rangle = \frac{|01\rangle + |10\rangle}{\sqrt{2}}, \tag{2.33}$$

$$|\beta_{10}\rangle = \frac{|00\rangle - |11\rangle}{\sqrt{2}}, \tag{2.34}$$

$$|\beta_{11}\rangle = \frac{|01\rangle - |10\rangle}{\sqrt{2}}. \tag{2.35}$$

If a state $|\psi_{AB}\rangle$ is entangled, then tracing out one of the two systems leads to a mixed state. If $|\psi_{AB}\rangle = |\psi_A\rangle \otimes |\psi_B\rangle$ is not entangled, then tracing out part A or part B of the space leads to $|\psi_B\rangle$ or $|\psi_A\rangle$, respectively. Recalling that $\text{Tr}(\rho^2) = 1$ if and only if ρ is a pure state, we have a simple formula for testing whether a state is entangled or not [12]. Therefore, the state $|\psi_{AB}\rangle$ is entangled if and only if

$$\text{Tr}(\text{Tr}_A(|\psi_{AB}\rangle \langle\psi_{AB}|)^2) < 1. \tag{2.36}$$

This procedure works well once the composite system $|\psi_{AB}\rangle$ is a quantum pure state.

Example 2.11. We want to use (2.36) to check the entangled state $|\beta_{00}\rangle$. First, we trace out the system B from the state $|\beta_{00}\rangle$:

$$\rho_A = \text{Tr}_B(|\beta_{00}\rangle \langle\beta_{00}|)$$

$$= \left[\frac{1}{2}|0\rangle \langle 0| + \frac{1}{2}|1\rangle \langle 1|\right].$$

Finally,

$$\text{Tr}(\rho_A^2) = \text{Tr}\left(\left[\frac{1}{2}|0\rangle \langle 0| + \frac{1}{2}|1\rangle \langle 1|\right]^2\right)$$

$$= \frac{1}{4} + \frac{1}{4}$$

$$< 1.$$

We investigate now the entanglement in the framework of quantum mixed states. Tensor products of mixed states, $\rho = \rho_1 \otimes \rho_2$, do not exhibit correlations, as do not the tensor products of pure states. A convex sum of different product states,

$$\rho = \sum_i p_i \rho_{1,i} \otimes \rho_{2,i}, \tag{2.37}$$

with $p_i > 0$ and $\sum_i p_i = 1$, will in general yield correlated measurement results, i.e., there are local observables a and b such that $\text{Tr}(\rho(a \otimes b)) \neq \text{Tr}(\rho(a \otimes \mathbb{1})) \, \text{Tr}(\rho(\mathbb{1} \otimes b)) = \text{Tr}_1 \rho_1 a \, \text{Tr}_2 \rho_2 b$. These correlations can be described in terms of the classical probabilities p_i and, therefore, they are considered classical. States like (2.37) are called *separable mixed states* [9].

In contrast, *mixed entangled states* are characterized by the non-existence of a decomposition into product states, as stated in the next definition.

Definition 2.9 (Mixed Entangled State [9]). A mixed state ρ is entangled if there are no local states $\rho_{1,i}$ and $\rho_{2,i}$, and non-negative weights p_i, such that ρ can be expressed as a convex mixture, i.e.,

$$\nexists \rho_{1,i}, \rho_{2,i}, p_i \geq 0 \text{ such that } \rho = \sum_i p_i \rho_{1,i} \otimes \rho_{2,i}. \tag{2.38}$$

Entanglement plays an important role in quantum information, communication, and computing. Perhaps, the most impressive application of entanglement is *teleportation*. Suppose that two physically separated parties, Alice and Bob, each takes one qubit of an EPR pair. Then, Alice can perform a teleportation of an arbitrary and unknown quantum state toward Bob by sending to him two classical bits of information. The quantum key distribution protocol proposed by Ekert [2], for instance, is based on this idea. A very didactic presentation of this protocol can be found in Fayngold and Fayngold [3].

2.6 Postulates of Quantum Mechanics

The concepts presented previously are organized in a framework of a workable physical theory, the so-called *postulates of quantum mechanics*. These postulates are a set of axioms that define how the theory operates [7]. According to the state-of-the-art knowledge, most of the rules in the universe can be traced back to these postulates and only a few effects seem to be an exception [5].

Frequently, the postulates of quantum mechanics are enunciated using the Dirac notation. Considering our purposes, we are going to enunciate these postulates using the density operators formalism, which is more convenient in quantum information and communication applications [10].

The first postulate tells us how physical states are represented in quantum mechanics. A quantum mechanical two-level system might be a single photon that can be found in one of the two distinct paths or a presence or absence of a photon in a particular location or path [6].

Postulate 2.1 (State Space of an Isolated Quantum System). *We associate with an isolated quantum system a complex vector space with inner product (i.e., a Hilbert space) known as* space state of the system. *This system is completely described by a density operator ρ, which is positive and has trace equal to 1, acting on the space state of the system. If the quantum system is in the state ρ_i with probability p_i, then the density operator of this system is $\sum_i p_i \rho_i$.*

The evolution of a closed quantum system is described by a unitary operator U. If the system is initially in the state $|\psi_i\rangle$ with probability p_i, after applying the operator U the state will be $U|\psi_i\rangle$ with probability p_i. Thus, the evolution of a quantum system according to the density operator framework is described by

$$\rho = \sum_i p_i |\psi_i\rangle \langle\psi_i| \xrightarrow{U} \sum_i p_i U |\psi_i\rangle \langle\psi_i| U^\dagger = U\rho U^\dagger. \qquad (2.39)$$

Postulate 2.2 (Evolution of Closed Quantum Systems). *The evolution of a closed quantum system is described by a unitary transformation. Therefore, the state of a quantum system ρ at time t_1 is associated with the state ρ' at time t_2 by means of a unitary operator U that depends only on t_1 and t_2:*

$$\rho' = U\rho U^\dagger. \qquad (2.40)$$

Measuring a quantum system that is in the state $|\psi\rangle$ seeks to obtain classical information about this state [11]. We can say that measurements connect the quantum and classical worlds; they are the only tools which allow taking a look at what happens in the quantum world [6]. Measuring the state of an unknown quantum system, in general, disturbs the state irreversibly. In those cases, there is no way to know or recover the state before the measurement. If the state was not disturbed, no new information about it is obtained [11]. Thus, measurements are obviously not reversible and therefore they represent the only exception under the unitary constraint.

The third postulate of quantum mechanics is synthesized as follows.

Postulate 2.3 (Quantum System Measurement). *Quantum measurements are described by a set of measurement operators $\{M_m\}$. These operators act on the space state of the quantum system being measured. The index m refers to the output that can occur at the measurement. If the state of the system prior to the measurement is ρ, then the probability of getting m at the measurement is*

$$p(m) = \mathrm{Tr}(M_m^\dagger M_m \rho). \tag{2.41}$$

Given that the output m occurred, the post-measurement state of the system will be

$$\rho' = \frac{M_m \rho M_m^\dagger}{\mathrm{Tr}(M_m^\dagger M_m \rho)}. \tag{2.42}$$

The set of measurement operators satisfies the completeness relation $\sum_m M_m^\dagger M_m = \mathbb{1}$.

The most common type of measurement in quantum mechanics is the projective measurement. This kind of measurement projects the system onto one of the eigen-subspaces of an observable and returns the corresponding eigenvalue. However, there exists a whole range of problems, such as pure state discrimination or joint measurement on several qubits, where it is more advantageous to use a general measurement procedure that tries to detect outcomes using a set of non-orthogonal operators. For such situations, a POVM measurement is adequate [3].

So far we have discussed the postulates for the case of a single system. If we want to study potentially useful quantum computing and communication applications, we need to understand how quantum mechanics works for systems composed of several qubits interacting with each other [6]. Entanglement, for instance, arises from composite quantum systems defined in the fourth postulate.

Postulate 2.4 (Composite Quantum Systems). *The state space of a composite quantum system is the tensor product of the space of states that compose it. If these systems are numbered from 1 to n, and the system i is in the state ρ_i, then the state of the composite system will be $\rho_1 \otimes \rho_2 \otimes \ldots \otimes \rho_n$.*

In a very ingenious way, Portugal says that the postulates of quantum mechanics presented previously can be understood as "game rules." If you break then, you are out of the game, i.e., you must respect them to create and understand quantum algorithms, protocols, etc. Considering the idea of game rules, the first postulate can be described as the arena where the game goes on. The second describes the dynamics of the game. The third describes the process of physical measurement. The fourth postulate describes how we adjoin various systems [11].

2.7 Further Reading

In this chapter we introduced an overview of some important quantum mechanics concepts. We presented the notion of qubits, evolution of quantum systems and projective and POVM measurements using the Dirac notation [1], widely known for simplifying the operations to be performed. We showed how entanglement represents non-trivial correlations between two or more quantum systems. Lastly, we introduced the density operators and enunciated the quantum mechanics postulates according to this framework.

Concepts presented in this chapter are an overview organized from many works in the literature: Williams [13], Nielsen and Chuang [10], Kaye et al. [6], Imre and Balazs [5], Hirvensalo [4], McMahon [7], Fayngold and Fayngold [3], among others. We kindly recommend these references for further reading.

References

1. Dirac P (1982) The principles of quantum mechanics, 4th edn. Oxford University Press, Oxford
2. Ekert A (1991) Quantum cryptography based on Bell's theorem. Phys Rev Lett 67:661–663
3. Fayngold M, Fayngold V (2012) Quantum mechanics and quantum information. Wiley, Singapore
4. Hirvensalo M (2004) Quantum computing. Springer, Berlin
5. Imre S, Balazs F (2005) Quantum computing and communications - an engineering approach. Wiley, Chichester
6. Kaye P, Laflamme R, Mosca M (2007) An introduction to quantum computing. Oxford University, Oxford
7. McMahon D (2008) Quantum computing explained, 1st edn. Wiley, New York
8. Mermin ND (2007) Quantum computer science – an introduction. Cambridge University Press, Cambridge
9. Mintert F, Viviescas C, Buchleitner A (2009) Basic concepts of entangled states. Springer, New York, pp 61–86
10. Nielsen MA, Chuang IL (2010) Quantum computation and quantum information. Cambridge University Press, Cambridge
11. Portugal R (2013) Quantum walks and search algorithms. Springer, New York
12. Vedral V (2006) Introduction to quantum information science. Oxford University Press, Oxford
13. Williams CP (2011) Explorations in quantum computing. 2nd edn. Springer, New York

Chapter 3
Fundamentals of Information Theory

The statistical theory of communication introduced by Claude Shannon answers fundamental questions about a system consisting of agents exchanging classical information through a classical communication channel [20]. The proposition of information theory was made in the electrical engineering domain due to the problems with telegraphs, radios, and telephones that the engineers faced in order to attend a growing demand of customers. Such practical challenges motivated the search for answers to fundamental questions such as how to define and quantify information, bounds to information compression and also strategies to exchange information securely [9]. Nowadays, information theory lies at the heart of modern technology, underpinning all communications, networking, and data storage systems.

The history of classical information theory began with Claude Shannon [28]. In this paper, he coined the essential terminology, and he stated and justified the main mathematical definitions, as well as the two fundamental theorems of information theory. The *noiseless coding theorem* quantifies the physical resources required to store the output of an information source; the second, called *channel coding theorem*, quantifies how much information it is possible to reliably transmit through a noisy communication channel [24].

We can say that information theory has two primary goals. The first is the development of fundamental theoretical limits on the achievable performance when communicating a given information source over a communication channel and using a specified coding scheme. The second goal is the development of coding schemes that provide reasonably good performance in comparison with the optimal theoretical limits established by the theory. [14].

While *classical information theory* refers to a mathematical framework for modeling the manipulation and transmission of classical information, *quantum information theory* studies fundamental problems related to the transmission of classical and quantum information over quantum communication channels. Quantum information theory promises to lead to a deeper understanding of fundamental

© Springer International Publishing Switzerland 2016
E.B. Guedes et al., *Quantum Zero-Error Information Theory*,
DOI 10.1007/978-3-319-42794-2_3

properties of nature and, at the same time, support new and exciting applications [20]. Quantum information is a fascinating topic precisely because it shows that the laws of information processing are actually dependent on the laws of physics [36].

This chapter introduces some elementary concepts regarding information theory. We start with classical information theory, aiming at introducing the two important theorems of Claude Shannon. First, we present entropy and other measures of information in Sect. 3.1.1. Then, we discuss in Sect. 3.1.2 a very important quantity in classical information theory, the capacity of a noisy classical channel. In the second part of the chapter, we introduce quantum information theory. We start with von Neumann entropy and other measures of information in Sect. 3.2.1. Section 3.2.2 introduces the definitions regarding quantum channels, including the Choi-Jamiołkowski isomorphism; the accessible information, Holevo quantity and the first type of quantum capacity are discussed in Sect. 3.2.3. The classical capacity of quantum channels is presented in Sect. 3.2.4. Finally, Sect. 3.2.5 discusses other capacities of quantum channels.

3.1 Classical Information Theory

In 1948, while working at Bell Labs, Claude E. Shannon developed the first successful theory of information, where information is modeled as events which occur with certain probabilities [36]. Besides defining precisely what is information and how to measure it, Shannon proved the existence of codes that allow communication with a negligible probability of error, since the transmission rate is below a certain parameter, called the *channel capacity* [9].

According to information theory, an *information source* can be modeled as a physical device that outputs letters from some fixed alphabet, each letter with a given probability. Let $\mathcal{X}$ denote a source alphabet that consists of messages, say $x_1, x_2, \ldots$, with probabilities $p(x_1), p(x_2), \ldots$, satisfying

$$p(x) \geq 0, \forall x \in \mathcal{X}, \text{ and } \sum_{x \in \mathcal{X}} p(x) = 1. \qquad (3.1)$$

For our purposes, we are only interested in *stationary memoryless* information sources, where subsequent source outputs are not correlated; by stationary, we mean that the probability mass function of the information source does not change over time.

In information theory, the term "information" is not a quantity associated with individual messages, but rather characterizes the source of the messages. The point of characterizing the source is to discover what capacity is required in a communication channel to transmit all the messages the source produces [34].

Along this section we will depict the elementary concepts regarding classical information theory, presenting definitions and illustrative examples. We will focus our discussions on introducing the two main theorems proposed by Shannon [28] because they establish fundamental limits on data compression and reliable transmission over noisy classical channels.

3.1.1 Entropy and Other Measures of Information

Entropy is a measure of unpredictability of information content. The concept of entropy was introduced into thermodynamics in the nineteenth century. It was considered to be a measure of the extent to which a system was disordered. Shannon [28] introduced this concept into communication theory and it was then realized that entropy is a property of any stochastic system and the concept is now used widely in many fields, such as statistics, computing, information processing, among others [35].

In order to define entropy, let X be a discrete random variable with alphabet $\mathcal{X}$ and probability mass function given by $p(x) = \Pr[X = x], x \in \mathcal{X}$.

Definition 3.1 (Entropy). The binary entropy of X, denoted by $H(X)$, is defined as the expected amount of information gained from observing X:

$$H(X) = -\sum_{x \in \mathcal{X}} p(x) \log p(x). \tag{3.2}$$

The logarithm is taken in base 2 and we consider that $0 \log 0 \equiv 0$. Entropy is expressed in *bits*. As expected, the entropy is not related with the values that the random variable can assume, but rather with their probabilities. Furthermore, entropy is a positive function, i.e., $H(X) > 0$, for any X.

Example 3.1. Let X be a random variable related to an unfair dice whose even values can appear twice more often than the odd values. This random variable has alphabet $\mathcal{X} = \{1, \ldots, 6\}$ and probability mass function $p(1) = p(3) = p(5) = \frac{1}{9}$ and $p(2) = p(4) = p(6) = \frac{2}{9}$. The entropy of X is given by:

$$
\begin{aligned}
H(X) &= -\sum_{x \in X} p(x) \log p(x) \\
&= -(p(1) \log p(1) + \ldots + p(6) \log p(6)) \\
&= -\left(3 \cdot \frac{1}{9} \log \frac{1}{9} + 3 \cdot \frac{2}{9} \log \frac{2}{9}\right) \\
&\approx -\left[3 \cdot \frac{1}{9} \cdot (-3.1699) + 3 \cdot \frac{2}{9} \cdot (-2.1699)\right] \\
&\approx 2.5032 \text{ bits.}
\end{aligned}
$$

This entropy value means that, on average, it is necessary 2.5032 binary questions in order to discover the value of a realization of X. It is interesting to notice that this result is in sharp contrast with the entropy of a fair dice Y, $H(Y) \approx 2.5849$ bits. Given that some results of the random variable X are more likely to occur, the average number of binary questions is reduced, i.e., the uncertainty regarding X is lower than the uncertainty of Y.

Fig. 3.1 Binary entropy
function

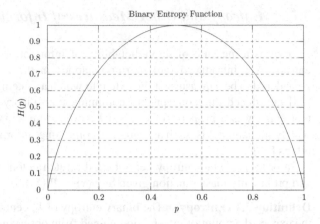

A special case for the entropy happens when the random variable has only two possible outcomes. For such situations, we define the *binary entropy function* as follows.

Definition 3.2 (Binary Entropy Function). Let X be a binary random variable with alphabet $\mathcal{X} = \{x_1, x_2\}$, and probability mass function $p(x_1) = p$ and $p(x_2) = 1 - p$. We denote by $H(p)$ the binary entropy function:

$$H(p) = H(X) = -p \log p - (1 - p) \log(1 - p), \tag{3.3}$$

recalling that $p \in [0, 1]$.

A graphic for the binary entropy function versus p is showed in Fig. 3.1. It is easy to see that

1. $H(p) > 0$ for $0 < p < 1$.
2. $H(p)$ is symmetric about $p = 0.5$.
3. $\lim_{\to 0} H(p) = \lim_{\to 1} H(p) = 0$.
4. $H(p)$ is increasing for $0 < p < 0.5$ and decreasing for $0.5 < p < 1$ and has a maximum for $p = 0.5$.
5. The binary entropy is a concave function of p.

Shannon entropy can be used to define other measures of information that capture relationships between two random variables X and Y. We define along this section the joint entropy, the conditional entropy, the relative entropy, and the mutual information.

The joint entropy is an extension of the entropy to a pair of random variables. It is defined as follows.

Definition 3.3 (Joint Entropy). The joint entropy $H(X, Y)$ of a pair of discrete random variables (X, Y) with joint distribution $p(x, y)$ is defined by

$$H(X, Y) = -\sum_{x \in \mathcal{X}} \sum_{y \in \mathcal{Y}} p(x, y) \log p(x, y). \tag{3.4}$$

Table 3.1 Joint distributions
of the random variables X
and Y

	y_1	y_2
x_1	0.3	0.1
x_2	0.1	0.5

Example 3.2. Let X and Y be two random variables with joint distribution shown in Table 3.1. The joint entropy $H(X, Y)$ is given by

$$H(X, Y) = -\sum_{x \in \mathcal{X}} \sum_{y \in \mathcal{Y}} p(x, y) \log p(x, y)$$

$$= -[p(x_1, y_1) \log p(x_1, y_1) + p(x_1, y_2) \log p(x_1, y_2) +$$

$$+ p(x_2, y_1) \log p(x_2, y_1) + p(x_2, y_2) \log p(x_2, y_2)]$$

$$= -[0.3 \cdot (-1.73) + 0.1 \cdot (-3.32) + 0.1 \cdot (-3.32) + 0.5 \cdot (-1)]$$

$$= 1.683 \text{ bits.}$$

The *conditional entropy* measures the information gained from learning the outcome of X given that Y is known.

Definition 3.4 (Conditional Entropy). If $(X, Y) \sim p(x, y)$, the conditional entropy $H(Y|X)$ is defined as

$$H(X|Y) = \sum_{x \in \mathcal{X}} p(x) H(Y|X = x)$$

$$= -\sum_{x \in \mathcal{X}} p(x) \sum_{y \in \mathcal{Y}} p(y|x) \log p(y|x)$$

$$= -\sum_{x \in \mathcal{X}} \sum_{y \in \mathcal{Y}} p(x, y) \log p(y|x), \tag{3.5}$$

where $p(y|x) = p(x, y)/p(x)$.

Example 3.3. Consider the random variables X and Y presented in Example 3.2 with probabilities given in Table 3.1. It follows that $H(Y|X)$ is given by

$$H(Y|X) = \sum_{x \in \mathcal{X}} p(x) H(Y|X = x)$$

$$= 0.4 H(0.75) + 0.6 H(0.1667)$$

$$\approx 0.4 \cdot 0.81 + 0.6 \cdot 0.648$$

$$= 0.712 \text{ bits.}$$

The conditional entropy $H(X|Y)$ is

$$H(X|Y) = \sum_{y \in \mathcal{Y}} p(y) H(X|Y = y)$$

$$= 0.4H(0.75) + 0.6H(0.1667)$$

$$\approx 0.712 \text{ bits.}$$

Although $H(Y|X)$ and $H(X|Y)$ are equal, this is not always the case.

The joint and conditional entropies can be used to build a relationship called the *chain rule*, given by

$$H(X, Y) = H(X) + H(X|Y), \tag{3.6}$$

which can be proved by applying the definition of $p(x|y)$.

Example 3.4. To check the chain rule, consider again the two random variables X and Y of Examples 3.2 and 3.3. We have shown that $H(X, Y) = 1.683$ bits and $H(Y|X) = 0.712$ bits. Considering that $H(X) = H(0.4) = 0.971$ bits, then $H(X, Y) = 0.971 + 0.712 = 1.683$, as stated by the chain rule.

Definition 3.5 (Relative Entropy). The relative entropy of two random variables X and Y is defined as follows:

$$H(X||Y) = -\sum_{x \in \mathcal{X}} \sum_{y \in \mathcal{Y}} p(x) \log p(y) - H(X)$$

$$= \sum_{x \in \mathcal{X}} \sum_{y \in \mathcal{Y}} p(x) \log \frac{p(x)}{p(y)}. \tag{3.7}$$

The *relative entropy* represents the difference between the expected information obtained from Y, given that they are distributed according to X, i.e., according to

$$\sum_{x,y} p(x) \log p(y),$$

and the expected information obtained from X, $H(X)$ [36].

Example 3.5. Let X be a random variable corresponding to the unfair dice described previously in Example 3.1, and let Y be a random variable corresponding to a fair dice. The relative entropy among X and Y is given by

$$H(X||Y) = \sum_{x \in \mathcal{X}} \sum_{y \in \mathcal{Y}} p(x) \log \frac{p(x)}{p(y)}$$

$$= 3 \cdot \left(\frac{1}{9} \sum_{y \in \mathcal{Y}} \log \frac{1/9}{p(y)} \right) + 3 \cdot \left(\frac{3}{9} \sum_{y \in \mathcal{Y}} \log \frac{3/9}{p(y)} \right)$$

$$= 3 \cdot \frac{1}{9} \cdot 6 \log \frac{1/9}{1/6} + 3 \cdot \frac{2}{9} \cdot 6 \log \frac{2/9}{1/6}$$

$$\approx -1.16 + 1.660$$

$$\approx 0.5 \text{ bits.}$$

We now introduce *mutual information*, which is a measure of the amount of information that one random variable contains about another random variable. The mutual information can also be understood as the reduction in the uncertainty of one random variable due to the knowledge of the other [9].

Definition 3.6 (Mutual Information). Consider two random variables X and Y with a joint probability mass function $p(x, y)$ and marginal probability mass functions $p(x)$ and $p(y)$. The mutual information $I(X; Y)$ is the relative entropy between the joint distribution and the product distribution $p(x)p(y)$:

$$I(X; Y) = \sum_{x \in \mathcal{X}} \sum_{y \in \mathcal{Y}} p(x, y) \log \frac{p(x, y)}{p(x)p(y)}$$

$$= H(X) - H(X|Y). \qquad (3.8)$$

If the variables X and Y are independent, then the mutual information $I(X; Y)$ is zero. If the random variables are completely correlated, then the mutual information between them is the information contained in X. More formally, if there exists a bijection f such that $P(X = x) = P(Y = f(x))$, then $I(X; Y) = H(X) = H(Y)$ [36].

Example 3.6. Consider the random variables X and Y from Example 3.4. The mutual information between them is given by

$$I(X; Y) = H(X) - H(X|Y)$$

$$= 0.971 - 0.712$$

$$= 0.259 \text{ bits.}$$

The Venn diagram in Fig. 3.2 illustrates the relationship between entropies, joint entropy, conditional entropies and mutual information between two random variables X and Y.

Fig. 3.2 Venn diagram of entropy and other measure functions and their relationship

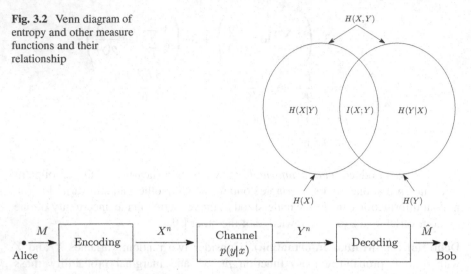

Fig. 3.3 Generic communication model

3.1.2 Capacity of a Noisy Classical Channel

Information does not only have to be used or stored, but it also has to be transmitted. In communication systems a sender converts (encodes) a message to a form suitable for transmission through a communication medium, be it a fiberoptic channel, satellite link, or radio signal through space. The receiver then detects the transmitted signal and decodes it back to the original message [35].

When two parties communicate, each one can influence the physical state of the other through some physical process. The precise nature of the parties and of the signaling process can be very different, thus it is necessary to consider an abstract model of communication [20].

Considering the importance of such abstraction, we follow Shannon's original work and consider a generic communication model, as shown in Fig. 3.3. In this model, a sender Alice wants to send a message M to a receiver Bob.

Alice is the *source* of the messages. She owns a predefined set of messages $M = \{M_1, M_2, \ldots, M_m\}$, and she can pick each one with a certain probability $p(M_1)$, $p(M_2), \ldots, p(M_m)$ that does not change over time. The first step to send the message M is to map it into symbols from the channel alphabet, denoted by $\mathcal{X}$, resulting in a *codeword* X^n of length n.

The *channel* can be any physical medium such as a telephone wire, the internet, a hard disk, an optic fiber, among others. In the real world, the data which is transferred through the channel may be subject to distortion, noise, and interference. Definition 3.7 formalizes the idea of a communication channel.

Definition 3.7 (Discrete Channel). A discrete channel is a triple $(\mathcal{X}, \mathcal{Y}, P)$, where $\mathcal{X} = \{x_1, x_2, \ldots, x_r\}$ is the input alphabet, $\mathcal{Y} = \{y_1, y_2, \ldots, y_s\}$ is the output alphabet and P denotes a probability transition matrix. The probability $P_{y,x} = p(y|x)$ denotes the probability of obtaining an output y when the channel's input was x.

The probability transition matrix has the form:

$$P = \begin{bmatrix} p(y_1|x_1)\ p(y_2|x_1) \cdots p(y_s|x_1) \\ p(y_1|x_2)\ p(y_2|x_2) \cdots p(y_s|x_2) \\ \vdots \qquad\qquad \ddots \\ p(y_1|x_r)\ p(y_2|x_r) \cdots p(y_s|x_r) \end{bmatrix}. \tag{3.9}$$

From the channel matrix, it is possible to see that

- Each row of P contains the probabilities of all possible outputs from the same input to the channel.
- Each column of P contains the probabilities of all possible inputs to a particular output from the channel.
- If we transmit the symbol x_i, then we must receive an output symbol with probability 1, that is:

$$\sum_{j=1}^{s} p(y_j|x_i) = 1, \text{ for } i = 1, 2, \ldots, r. \tag{3.10}$$

For a given discrete channel, we can construct a graph as follows. Each symbol in $\mathcal{X}$ corresponds to a vertex in the graph, labeled with the corresponding symbol; the same for the symbols in $\mathcal{Y}$. If $p(y_j|x_i) > 0$, there is an edge directed from x_i to y_j, labeled with $p(y_j|x_i)$. An example of such graph is illustrated in Fig. 3.4, where only the edges from x_1 and their labels are emphasized.

A widely adopted model of channel is the *discrete memoryless channel without feedback*, or *discrete memoryless channel* (DMC), for short. In this kind of channel, the behavior and the effect of the noise at time t will not depend on the behavior of the channel or the effect of noise at any previous time $t - 1, t - 2, \ldots$. This kind of channel is defined as follows.

Fig. 3.4 Graph of a discrete channel

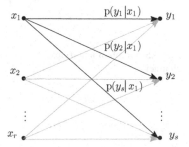

Definition 3.8 (Discrete Memoryless Channel). A discrete memoryless channel is a discrete channel whose transition probabilities can be factored in the following way:

$$p(y^N|x^N) = \prod_{i=1}^{N} p(y_i|x_i). \tag{3.11}$$

Example 3.7 (Noiseless Channel). Consider a DMC with binary input and output alphabets, labeled $\{0, 1\}$. The channel is said to be noiseless if there is no transmission errors. The probability transition matrix is given by

$$P = \begin{bmatrix} p(y=0|x=0) = 1 & p(y=1|x=0) = 0 \\ p(y=0|x=1) = 0 & p(y=1|x=1) = 1 \end{bmatrix},$$

and the graph corresponding to this channel is shown in Fig. 3.5.

Shannon verified an entropy-related bound on how good compression can be for a given source by using the notion of a noiseless channel.

Theorem 3.1 (Noiseless Coding Theorem [28]). *Let a classical source S emit symbols from an alphabet according to a given probability distribution. For n sufficiently large, a message sent of length n over a noiseless channel can be compressed without loss of information to a minimum of $H(S) \times n$ bits.*

Shannon's theorem establishes a bound for lossless compression algorithms, it does not provide us with one. Huffman's algorithm is the most famous and basic compression algorithm [41].

Example 3.8 (Binary Symmetric Channel). Consider a channel with $\{0, 1\}$ as input and output alphabets. Suppose that the channel introduces a bit flip 5 % of times. The graph of this channel, called *Binary Symmetric Channel* (BSC), is showed in Fig. 3.6. The channel matrix is given by

$$P = \begin{bmatrix} 0.95 & 0.05 \\ 0.05 & 0.95 \end{bmatrix}.$$

Fig. 3.5 Graph of a noiseless channel

0   1   0

1   1   1

Fig. 3.6 Graph of a bit-flip channel

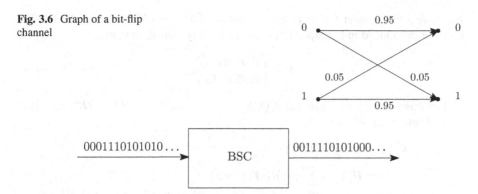

Fig. 3.7 Example of an input and output sequences from a binary symmetric channel

The channel is said to be symmetric because $p(y = 0|x = 0) = p(y = 1|x = 1)$ and $p(y = 1|x = 0) = p(y = 0|x = 1)$. An example of input and output sequences from this channel is shown in Fig. 3.7.

When a message of index M is transmitted by Alice using a codeword of length n, Bob guesses the index M of the transmitted message using an appropriate decoding rule $\hat{M} = g(Y^n)$. Bob makes a mistake if $\hat{M}$ is different from the index M that was transmitted. To formalize such scenario, we introduce some definitions.

Definition 3.9 ((m, n) Code). An (m, n) code for a DMC $(\mathcal{X}, \mathcal{Y}, P)$ consists of a set of messages $M = \{M_1, M_2, \ldots, M_m\}$, an encoder $f : M \to \mathcal{X}^n$, and a decoder $g : Y^n \to M$. The rate of this (m, n) code is

$$R = \frac{\log |M|}{n} \text{ bits per transmission.} \quad (3.12)$$

As mentioned before, the channel can introduce errors, causing the received message to be different from the original message. Despite the errors, we are interested in the maximum amount of information that can be transmitted per channel use. This quantity is defined as the *capacity* of the channel.

Definition 3.10 (Ordinary Channel Capacity). The maximum average mutual information, $I(X, Y)$, in any single use of a channel defines the channel capacity. Mathematically, the channel capacity is defined as

$$C = \max_{P_X} I(X; Y), \quad (3.13)$$

where I denotes the mutual information, X and Y are random variables representing the channel input and output, respectively. The maximum is taken over all possible input probability distributions.

Example 3.9 (Channel Capacity of the Bit-Flip Channel [23]). Recall the bit-flip channel introduced in Example 3.8, with probability transition matrix:

$$P = \begin{bmatrix} 0.95 & 0.05 \\ 0.05 & 0.95 \end{bmatrix}.$$

For this channel, we have that $H(Y|X = 0) = H(Y|X = 1) = H(0.05)$. The mutual information is given by

$$
\begin{aligned}
I(X; Y) &= H(Y) - H(Y|X) \\
&= H(Y) - \sum_{x \in \mathcal{X}} p(x) H(Y|X = x) \\
&= H(Y) - [p(x = 0)H(Y|X = 0) + p(x = 1)H(Y|X = 1)] \\
&= H(Y) - H(0.05) \\
&\leq 1 - H(0.05) \qquad\qquad\qquad\qquad\qquad\qquad (3.14) \\
&\leq 1 - 0.2863 \\
&\leq 0.7137 \text{ bits,}
\end{aligned}
$$

where (3.14) follows because Y is a binary random variable. Since equality in (3.14) is attained if Y is uniform, which will hold if input X is uniform, we conclude that the capacity of this channel is given by

$$C = 1 - H(0.05) = 0.7137 \text{ bits.}$$

Therefore, the capacity is achieving when X assumes a uniform distribution, $p(x = 0) = P(x = 1) = 1/2$.

The channel capacity has the following properties [9]:

- $C \geq 0$, since $I(X; Y) \geq 0$.
- $C \leq \log |\mathcal{X}|$, since $C = \max I(X; Y) \leq \max H(X) = \log |\mathcal{X}|$.
- $C \leq \log |\mathcal{Y}|$ for the same reason.
- $I(X; Y)$ is a continuous function of $p(x)$.
- $I(X; Y)$ is a concave function of $p(x)$.

Example 3.10 (Binary Erasure Channel). In a binary erasure channel (BEC), whose graph representation is shown in Fig. 3.8, the input alphabet is $\mathcal{X} = \{0, 1\}$ and the output alphabet is $\mathcal{Y} = \{0, 1, ?\}$, where ? means that the input bit was lost. It is important to emphasize that the receiver knows when the input is lost, i.e., when neither 0 nor 1 was actually received. This channel matrix is given by

$$P = \begin{bmatrix} 1 - \gamma & \gamma \\ \gamma & 1 - \gamma \end{bmatrix}, \qquad\qquad\qquad\qquad (3.15)$$

where γ is the *erasure probability*.

Fig. 3.8 Graph of a binary
erasure channel

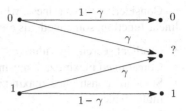

The binary erasure channel capacity is given as follows:

$$C = \max_P I(X; Y)$$

$$= \max_P (H(Y) - H(Y|X))$$

$$= \max_P H(Y) - H(\gamma).$$

The reader may think that the maximum of $H(Y)$ would be $\log 3$, since the output
alphabet contains three symbols. However, we cannot achieve this because of the
input distribution. Let E be the event $\{Y = ?\}$. Then,

$$H(Y) = H(Y, E) = H(E) + H(Y|E).$$

Let $\Pr(x = 1) = \pi$. The entropy of Y will be

$$H(Y) = H((1 - \pi)(1 - \gamma), \gamma, \pi(1 - \gamma))$$

$$= H(\gamma) + (1 - \gamma)H(\pi).$$

Using these results, the BEC capacity becomes

$$C = \max_P H(Y) - H(\gamma)$$

$$= \max_\pi (1 - \gamma)H(\pi) + H(\gamma) - H(\gamma)$$

$$= \max_\pi (1 - \gamma)H(\pi)$$

$$= 1 - \gamma,$$

where the capacity is achieved when $\pi = \frac{1}{2}$. This result is intuitively reasonable:
since a proportion of γ bits are lost in the channel, it is possible to recover (at most)
a proportion of $(1 - \gamma)$ bits. Hence, the capacity is $(1 - \gamma)$ bits per channel use.

Calculating C is not an easy task because it involves maximization of $I(X; Y)$
over $r = \log |\mathcal{X}|$ independent variables, subject to two constraints:

1. $p(x) \geq 0, \forall x \in \mathcal{X}.$
2. $\sum_{x \in \mathcal{X}} p(x) = 1.$

Considering this hardness, which is due to a constrained maximization of a non-linear function, some methods are suggested by the proper literature [9, 35]:

- Gradient search algorithms.
- Constrained maximization using calculus and the Kuhn-Tucker conditions.
- Standard constrained maximization techniques like the method of Lagrangian multipliers.
- Iterative algorithms developed by Arimoto [1] and Blahut [7].
- Derivation for special cases.

In general, there is no closed-form solution for the capacity. But for many simple channels it is possible to calculate the capacity using properties such as symmetry [9].

The channel capacity measures the amount of information that can be carried over the channel; in fact, it characterizes the maximal amount of transmission rate for reliable communication. Prior to the mid-1940s people believed that transmitted data subject to noise corruption can never be perfectly recovered unless the transmission rate approaches zero. Shannon's landmark work [28] disproved this thinking and established the well-known *Channel Coding Theorem*: as long as the transmission rate (in bits per channel use) is below (but can be arbitrarily close to) the channel capacity, the error can be made smaller than any given number (which we term arbitrarily small) by some properly designed coding scheme [23].

Definition 3.11 (Average Error Probability). Let

$$\lambda_i = \Pr[g(Y^n) \neq i | X^n = f(i)] \tag{3.16}$$

be the conditional probability that the receiver Bob makes a wrong guess, given that the i-th codeword is sent. The average error probability $\lambda^{(n)}$ for an (m, n) coding scheme is defined as

$$\lambda^{(n)} = \frac{1}{M} \sum_{i=1}^{M} \lambda_i. \tag{3.17}$$

Theorem 3.2 (Channel Coding Theorem [28]). *For a discrete memoryless channel, it is possible to transmit messages with an arbitrarily small error probability if the communication rate R is below the channel capacity C. Specifically, for every rate $R < C$ there exists a sequence of $(2^{nR}, n)$ coding schemes with average error probability $\lambda^{(n)} \to 0$ as $n \to \infty$.*

This theorem reconciles two competing parameters for a noisy channel: (1) a high transmission rate; and (2) a low error probability [8]. Some ideas that follow from this theorem are: allowing an arbitrarily small but nonzero probability of error; using the channel many times in succession, so that the law of large numbers comes into effect; and calculating the average of the probability of error over a random choice of codebooks, which symmetrizes the probability, and which can then be

used to show the existence of at least one good code [9]. We refer to the book of
Desurvire [12, Sect. 13.3] for a complete detailed proof of this theorem, presented
with three different approaches.

Although the channel coding theorem guarantees the existence of good coding
schemes with arbitrarily small error probability for long n, it does not provide a way
of constructing the best coding schemes. Ever since Shannon's original findings,
researchers have tried to develop practical coding schemes that are easy to encode
and decode [23].

3.2 Quantum Information Theory

Quantum information theory studies fundamental problems related to transmission
of classical and quantum information over quantum communication channels,
such as the entropy of quantum systems, the classical and quantum capacities
of quantum channels, the effect of the noise, fidelity, and optimal information
encoding. Quantum information theory promises to lead to a deeper understanding
of fundamental properties of nature and, at the same time, support new and exciting
applications [20].

Quantum information theory intersects two of the great sciences of the twentieth
century: the quantum theory and information theory. It was really only a matter of
time before physicists, mathematicians, computer scientists, and engineers began
to consider the convergence of the two subjects, as quantum theory was essentially
established by 1926 and information theory by 1948 [39].

We can think of classical information theory as a subset of quantum information
theory, where we are restricted to orthogonal states. In this view, there is no division
between the classical and quantum worlds. However, since quantum information
seems to be broader than classical information, a question that arises is how
quantum information can be characterized? It turns out, however, that quantum
information can be quantified in the same way as classical information using
Shannon's prescription [36].

The second part of the chapter will depict important concepts regarding quantum
information theory, making, when possible, analogies with the classical domain.
After some elementary definitions, we will focus our discussions to quantum
channels and their capacities. In quantum information theory, quantum channels
can be used in many different ways: they can transmit classical information, private
classical information, or quantum information. A quantum channel can be used
alone, with shared entanglement, or even together with other classical and quantum
channels.

3.2.1 von Neumann Entropy and Other Measures of Information

The Shannon entropy measures the uncertainty associated with a classical probability distribution. Quantum states, on the other hand, are described by density operators, instead of probability distributions [24]. Considering such difference, John von Neumann realized that quantum mechanics required a definition of entropy covering pure and also mixed states [20]. He introduced such mathematical formalism for quantum mechanics in a book published in 1932 [38].

Firstly, we need to define a quantum analogue of a classical source. A quantum source outputs d quantum states $|\psi_1\rangle, |\psi_2\rangle, \ldots, |\psi_d\rangle$ with corresponding probabilities $p_1, p_2, \ldots, p_d$. States $|\psi_i\rangle$ are not restricted to be orthogonal. The quantum source can be characterized by a density operator ρ given by

$$\rho = \sum_{i=1}^{d} p_i |\psi_i\rangle \langle\psi_i|, \tag{3.18}$$

where $\sum_{i=1}^{d} p_i = 1$.

We are now ready to define the von Neumann entropy.

Definition 3.12 (von Neumann Entropy). The von Neumann entropy of a quantum state ρ is

$$S(\rho) = -\operatorname{Tr}(\rho \log \rho). \tag{3.19}$$

The logarithm is taken on base 2. The logarithm of ρ can be calculated by taking its spectral decomposition $\rho = \sum_i \lambda_i |\varphi_i\rangle \langle\varphi_i|$, where $\log \rho = \sum_i \log(\lambda_i) |\varphi_i\rangle \langle\varphi_i|$. Because λ_i are the eigenvalues of ρ and $\{|\varphi_i\rangle\}$ forms an orthonormal set, the von Neumann entropy can be written as

$$S(\rho) = -\operatorname{Tr}\left[\sum_i |\varphi_i\rangle \langle\varphi_i| \sum_j \log \lambda_j |\varphi_j\rangle\langle\varphi_j|\right]$$

$$= -\operatorname{Tr}\left[\sum_i \lambda_i \log \lambda_i |\varphi_i\rangle \langle\varphi_i|\right]$$

$$= -\sum_i \lambda_i \log \lambda_i, \tag{3.20}$$

with $0 \log 0 \equiv 1$.

From the definition, it is straightforward to see that the von Neumann entropy of a density matrix ρ is the Shannon entropy of its eigenvalues λ_i. The von Neumann entropy can be understood as a measure of the mixedness of ρ. This measure is bounded by $0 \leq S(\rho) \leq \log d$; when $S(\rho) = 0$, ρ is a pure state. The maximum $\log d$ corresponds to the quantum state $\rho = \mathbb{1}_d/d$. In this case, we have a maximum ignorance about the state of the system. The state $\rho = \mathbb{1}_d/d$ characterizes a *completely depolarized system* [22].

Example 3.11 (von Neumann Entropy of a Pure State). Let ρ be the density operator corresponding to the pure state of Example 2.7:

$$\rho = \frac{1}{2} \begin{bmatrix} 1 & 1 \\ 1 & 1 \end{bmatrix}.$$

The eigenvalues of ρ are $\{0, 1\}$. Therefore, the von Neumann entropy of ρ is given by

$$S(\rho) = -0 \log 0 - 1 \log 1$$
$$= 0.$$

Example 3.12 (von Neumann Entropy of a Mixed State [21]). Let σ be a quantum state with the following density matrix

$$\sigma = \begin{bmatrix} \frac{1}{2} & \frac{1}{4} \\ \frac{1}{4} & \frac{1}{2} \end{bmatrix}.$$

The eigenvalues of σ are $\{\frac{3}{4}, \frac{1}{4}\}$. We can see that σ is a mixed state:

$$\sigma^2 = \begin{bmatrix} \frac{1}{2} & \frac{1}{4} \\ \frac{1}{4} & \frac{1}{2} \end{bmatrix} \begin{bmatrix} \frac{1}{2} & \frac{1}{4} \\ \frac{1}{4} & \frac{1}{2} \end{bmatrix} = \begin{bmatrix} \frac{5}{16} & \frac{1}{4} \\ \frac{1}{4} & \frac{5}{16} \end{bmatrix}$$

$$\Rightarrow \mathrm{Tr}(\sigma^2) = \frac{5}{16} + \frac{5}{16} = \frac{5}{8}.$$

Therefore, the von Neumann entropy of σ is

$$S(\sigma) = -\mathrm{Tr}\,\sigma \log \sigma$$
$$= -\frac{3}{4} \log \frac{3}{4} - \frac{1}{4} \log \frac{1}{4}$$
$$\approx 0.3112 + 0.5$$
$$\approx 0.8112.$$

Consider that ρ is a composite separable state, i.e., it has the form $\rho = \rho_1 \otimes \rho_2 \otimes \ldots \otimes \rho_n$. The entropy is *additive* in the following sense:

$$S(\rho) = S(\rho_1 \otimes \rho_2 \otimes \ldots \otimes \rho_n) = S(\rho_1) + S(\rho_2) + \ldots + S(\rho_n). \qquad (3.21)$$

The von Neumann entropy is *subadditive* [24, p. 515]. If ρ_1 and ρ_2 are reduced density matrices of a composite system ρ, the *subadditivity inequality* states that

$$S(\rho) \leq S(\rho_1) + S(\rho_2). \tag{3.22}$$

Example 3.13 (von Neumann Entropy of a Completely Mixed State [21]). Consider that Alice and Bob share the Bell state

$$|\beta_{10}\rangle = \frac{|00\rangle - |11\rangle}{\sqrt{2}},$$

where the first qubit belongs to Alice and the second belongs to Bob. A density operator that describes this state is

$$\rho = |\beta_{10}\rangle \langle\beta_{10}|$$
$$= \frac{|00\rangle \langle 00| - |00\rangle \langle 11| - |11\rangle \langle 00| + |11\rangle \langle 11|}{2}.$$

In matrix form, we have

$$\rho = \frac{1}{2} \begin{bmatrix} 1 & 0 & 0 & -1 \\ 0 & 0 & 0 & 0 \\ 0 & 0 & 0 & 0 \\ -1 & 0 & 0 & 1 \end{bmatrix}.$$

The eigenvalues of ρ are $\{1, 0, 0, 0\}$. We can quickly see that the entropy of ρ is $S(\rho) = -\log 1 = 0$.

If we trace out the first qubit, we get the reduced density matrix for Bob:

$$\rho_B = \mathrm{Tr}_A \rho = \frac{1}{2} \begin{bmatrix} 1 & 0 \\ 0 & 1 \end{bmatrix}.$$

This is a completely mixed state with entropy given by

$$S(\rho_B) = -\log \frac{1}{2} = 1.$$

We find a similar result for Alice if we make a partial trace on the second qubit. It is also possible to see that $S(\rho) \leq S(\rho_A) + S(\rho_B)$ is satisfied.

The von Neumann entropy is a measure of our ignorance about the quantum state, and plays a similar role for quantum states as the Shannon entropy does for classical random variables. From the point of view of the von Neumann entropy, a quantum state can be understood as a quantum information source.

Besides additivity and subadditivity, some other properties of the von Neumann entropy are

- **Purity**. If ρ is a pure state, then $S(\rho) = 0$.
- **Invariance**. If U is an unitary transformation, then $S(\rho) = S(U\rho U^\dagger)$ for any ρ.
- **Concavity**. Provided $p_i \geq 0$ and $\sum_i p_i = 1$, then $S\left(\sum_i p_i \cdot \rho_i\right) \geq \sum_i p_i S(\rho_i)$. This result shows that the less we know about how a state is prepared the greater its von Neumann entropy.
- **Boundness**. $S(\rho) \leq H(\{p_i\})$ for an ensemble of quantum states $|\psi_i\rangle$ with probabilities p_i, and $\rho = \sum_i p_i |\psi_i\rangle$. The von Neumann entropy of the density matrix is never greater than the Shannon entropy of the corresponding classical ensemble. Equality holds when the quantum states are pairwise orthogonal and hence unambiguously distinguishable.
- **Strong Subadditivity**. For two systems AB and BC having common subsystem B, the sum of the von Neumann entropies of their union and intersection is less than the sum of their von Neumann entropies, i.e., $S(\rho_{ABC}) + S(\rho_B) \leq S(\rho_{AB}) + S(\rho_{BC})$.
- **Arak-Lieb Inequality**. A bipartite state ρ_{AB} can be completely known (zero entropy) even though its parts are not, such as when $S(\rho_A) = S(\rho_B) \neq 0$. In other words, $S(\rho_{AB}) \geq |S(\rho_A) - S(\rho_B)|$ [40, Sect. 11.4.1].

Making an equivalence with the classical concepts presented previously, it is also possible to define a quantum version of the joint entropy, which refers to the quantum entropy of a combined system.

Definition 3.13 (Quantum Joint Entropy). Given a quantum system ρ_{AB} with two subsystems ρ_A and ρ_B, the quantum joint entropy of the combined system is

$$S(\rho_A, \rho_B) = S(\rho_{AB}) = -\operatorname{Tr} \rho_{AB} \log \rho_{AB}. \tag{3.23}$$

As we already know, the von Neumann entropy of a pure state is zero. When a pure state describes an entangled composite quantum system, the corresponding reduced density operator has positive von Neumann entropy. Thus, the joint entropy of an entangled quantum system can be negative. This contrasts with the classical joint entropy, which is never negative [36].

Definition 3.14 (Quantum Conditional Entropy). Given two subsystems A and B and their composite system AB, the quantum conditional entropy is defined as

$$S(\rho_A|\rho_B) = S(\rho_{AB}) - S(\rho_B). \tag{3.24}$$

This definition is equivalent to the classical chain rule (3.6). If the joint system ρ_{AB} is in a pure state, we have that $S(\rho_{AB}) = 0$. From (3.24), the conditional entropy becomes $S(\rho_A|\rho_B) = -S(\rho_B) \leq 0$. The same holds for $S(\rho_B|\rho_A) = -S(\rho_A) \leq 0$. Hence, the quantum conditional entropy can be negative, which is definitely a nonclassical feature [12].

The relative entropy for the quantum domain is also defined.

Definition 3.15 (Quantum Relative Entropy). Let ρ and σ be two density operators. The quantum relative entropy of ρ to σ is defined by

$$S(\rho||\sigma) = \mathrm{Tr}\,\rho \log \rho - \mathrm{Tr}\,\sigma \log \sigma. \tag{3.25}$$

The quantum relative entropy can sometimes be infinite, as can the classical relative entropy. In particular, the relative entropy is defined to be $+\infty$ if the kernel of σ, the vector space spanned by the eigenvectors of σ with zero eigenvalues, has non-trivial intersection with the support of ρ, the vector space spanned by the eigenvectors of ρ with non-zero eigenvalues. Otherwise, the quantum relative entropy is finite [24].

The relative entropy tells us how similar two density operators are. It is minimal when $\rho = \sigma$, in which case $S(\rho||\rho) = 0$. Other properties of the relative entropy are listed below:

- **Additivity.** $S(\sigma_1 \otimes \sigma_2||\rho_1 \otimes \rho_2) = S(\sigma_1||\rho_1) + S(\sigma_2||\rho_2)$. The relative entropy inherits additivity from the von Neumann entropy.
- **Non-Negativity.** From Klein's inequality, the quantum relative entropy is always non-negative, $S(\rho||\sigma) \geq 0$, with equality if and only if $\rho = \sigma$ [24].
- **Convexity.** $S(\lambda\sigma_1 + (1-\lambda)\sigma_2||\rho) \leq \lambda S(\sigma_1||\rho_1) + (1-\lambda)S(\sigma_2||\rho_2)$. This rule says that the relative entropy is convex, which means that mixing of physical states decreases the distance between them.
- **Invariance.** $S(U\rho U^\dagger||U\sigma U^\dagger) = S(\rho||\sigma)$. The quantum relative entropy is invariant under unitary transformation.
- **Partial Trace Decreases Distinguishability.** $S(\mathrm{Tr}_B\,\sigma|||\,\mathrm{Tr}_B\,\rho) \leq S(\sigma||\rho)$. The less information we have about two states, the less we can tell if there is any difference between them.
- **Donald's Inequality.** The average distance to σ equals the average distance to ρ plus the distance from ρ to σ, where ρ is the average state. In a mathematical notation, $\sum_k p_k S(\rho_k||\sigma) = \sum_k p_k S(\rho_k||\rho) + S(\rho||\sigma)$, where $\rho = \sum_k p_k \rho_k$ [36].

Definition 3.16 (Quantum Mutual Information). Let A and B be two subsystems of a larger quantum system AB. We define the quantum mutual information $S(\rho_A; \rho_B)$ as

$$S(\rho_A; \rho_B) = S(\rho_A) + S(\rho_B) - S(\rho_{AB}). \tag{3.26}$$

Analogously to the classical case, the quantum mutual information represents the measure of information correlation between two quantum subsystems A, B. If the composite system AB is not correlated, $\rho_{AB} = \rho_A \otimes \rho_B$, then $S(\rho_a; \rho_B)$ is zero. If the joint system is pure ($S\rho_{AB} = 0$ and $S(\rho_A) = S(\rho_B)$), then $S(\rho_A; \rho_B) = 2S(\rho_A) = 2S(\rho_B)$.

The Venn diagram in Fig. 3.9 illustrates the relationship between the von Neumann entropy, quantum joint entropy, quantum conditional entropies and mutual information between two systems ρ_A and ρ_B.

Fig. 3.9 Venn diagram of
von Neumann entropy and
other quantum measure
functions and their
relationship

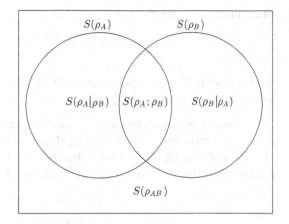

$$S(\rho_A) \qquad S(\rho_B)$$

$$S(\rho_A|\rho_B) \quad S(\rho_A;\rho_B) \quad S(\rho_B|\rho_A)$$

$$S(\rho_{AB})$$

3.2.2   Quantum Channels

The time evolution of a closed quantum system ρ is completely described by
unitary operators. If the system remains closed, it is always possible to return to
the initial system state. Suppose that a closed quantum system interacts in some
way with an open system, here called *environment*. Additionally, suppose that after
the interaction the system becomes closed again. We denote by $\mathcal{E}(\rho)$ the state of the
system after interaction.

In general, the final state $\mathcal{E}(\rho)$ cannot be related by a unitary transformation to the
initial state ρ. The formalism used to deal with such situation is known as *quantum
operations*. A quantum operation is a map $\mathcal{E}$ from the set of operators of the input
space $\mathcal{H}_1$ to the output state space $\mathcal{H}_2$ with the following properties[1] [22, 24]:

- **Property 1.** $\mathrm{Tr}[\mathcal{E}(\rho)]$ is the probability that the process represented by $\mathcal{E}$ occurs,
  when ρ is the initial state. Thus $0 \leq \mathrm{Tr}[\mathcal{E}(\rho)] \leq 1$ for any state ρ.
- **Property 2.** $\mathcal{E}$ is a convex-linear map on the set of density operators, that is, for
  probabilities $\{p_i\}$:

$$\mathcal{E}\left(\sum_i p_i \rho_i\right) = \sum_i p_i \mathcal{E}(\rho_i). \tag{3.27}$$

- **Property 3.** $\mathcal{E}$ is a completely positive map. If $\mathcal{E}$ maps density operators of system
  $\mathcal{H}_1$ to density operator of system $\mathcal{H}_2$, then $\mathcal{E}(A)$ must be positive for any positive
  operator A. Furthermore, $(\mathbb{1} \otimes \mathcal{E})(B)$ must be positive for any positive operator
  B on a composite system $R\mathcal{H}_1$, where $\mathbb{1}$ denotes the identity map on R.

The proof of the next theorem can be found in Nielsen and Chuang [24, p. 368].

---

[1]For the sake of simplicity, we consider $\mathcal{H}_1 = \mathcal{H}_2 = \mathcal{H}$.

Theorem 3.3. *A map $\mathcal{E}$ satisfies properties 1, 2, and 3 if*

$$\mathcal{E}(\rho) = \sum_i E_i \rho E_i^\dagger, \tag{3.28}$$

for some set of Kraus operators $\{E_i\}$, which maps the input Hilbert space to the output Hilbert space, and $\sum_i E_i^\dagger E_i \le \mathbb{1}$.

To model a *quantum channel*, it is required that the map $\mathcal{E}$ takes a valid density operator ρ into another valid one $\mathcal{E}(\rho)$. Hence, quantum channels belong to a class of maps called completely positive trace-preserving maps, which are completely positive maps that preserve the trace of operators, i.e.,

$$\begin{aligned}
1 &= \operatorname{Tr} \rho \\
&= \operatorname{Tr} \mathcal{E}(\rho) \\
&= \operatorname{Tr} \sum_i E_i \rho E_i^\dagger \\
&= \operatorname{Tr} \sum_i E_i^\dagger E_i \rho.
\end{aligned} \tag{3.29}$$

Since this relationship is valid for all ρ, then we must have

$$\sum_i E_i^\dagger E_i = \mathbb{1}. \tag{3.30}$$

Putting these concepts together, we present a formal definition of quantum channels.

Definition 3.17 (Quantum Channel). A quantum channel $\mathcal{E}$ is a trace-preserving completely positive map that acts on an input state ρ as follows:

$$\mathcal{E}(\rho) = \sum_i^m E_i \rho E_i^\dagger, \tag{3.31}$$

where $\{E_i\}_{i=1}^m$ is a set of Kraus operators (or operation elements) satisfying the completeness relation $\sum_{i=1}^m E_i^\dagger E_i = \mathbb{1}$. This way of defining quantum channels is known as the *operator-sum* formalism.

The simplest quantum channel is the identity channel, $\mathcal{E} \equiv \mathbb{1}$, which leaves intact the input quantum state. The identity channel is the quantum counterpart of the noiseless channel discussed in Example 3.7. For other quantum channels, errors and noise may occur.

Example 3.14 (Bit-Flip Quantum Channel). In a bit-flip quantum channel the state of a qubit is changed with probability $1 - p$ from $|0\rangle$ to $|1\rangle$ and vice versa. Its operation elements are

$$\left\{ \sqrt{p}\,\mathbb{1}, \sqrt{1-p}\,X \right\}.$$

This way, the operator-sum representation is

$$\mathcal{E}(\rho) = p\rho + (1-p)X\rho X.$$

Example 3.15 (Quantum Depolarizing Channel [24]). Consider a quantum depolarizing channel. In a 2-dimensional Hilbert space, this channel leaves a qubit intact with probability $1 - p$ and replaces the input state by a completely depolarized state $\mathbb{1}_2$ with probability p:

$$\mathcal{E}(\rho) = (1-p)\rho + p\frac{1}{2}\mathbb{1}_2.$$

The Kraus operators of this quantum channel are given by the following set:

$$\left\{ \sqrt{1 - \frac{3p}{4}}\mathbb{1}_2, \frac{\sqrt{p}}{2}X, \frac{\sqrt{p}}{2}Y, \frac{\sqrt{p}}{2}Z \right\}. \qquad (3.32)$$

Therefore, the operator-sum representation of the quantum depolarizing channel is

$$\mathcal{E}(\rho) = (1-p)\rho + \frac{p}{3}(X\rho X + Y\rho Y + Z\rho Z),$$

according to which the state ρ is left untouched with probability $1 - p$ and the operators X, Y, and Z are applied each with probability $p/3$.

The generalization of this quantum channel to an arbitrary dimension d is given by

$$\mathcal{E}(\rho) = (1-p)\rho + \frac{p}{d}\mathbb{1}_d. \qquad (3.33)$$

3.2.2.1 The Choi-Jamiołkowski Isomorphism

The use of operator-sum formalism to describe non-unitary evolutions of quantum system, as stated in Definition 3.17, is sufficient to understand most of the concepts presented in this book. However, there are many ways of representing trace-preserving completely positive (TPCP) quantum maps. The so-called Choi-Jamiołkowski isomorphism associates a linear operator with an Hermitian matrix in the following way. Consider a quantum channel $\mathcal{E}$, a TPCP map from the set of operators of the input Hilber space $\mathcal{H}_1$ to the set of operators in $\mathcal{H}_2$. For simplicity, consider $\mathcal{H} = \mathcal{H}_1 = \mathcal{H}_2$ and $\dim(\mathcal{H}) = d$. Let $|\omega\rangle_{AB} = \sum_i |i\rangle_A |i\rangle_B$ be a bipartite, full Schmidt-rank[2] state belonging to $\mathcal{H} \otimes \mathcal{H}$, where $|i\rangle$ stands for the computational

---

[2]The Schmidt decomposition of a bipartite quantum state $|\psi\rangle_{AB} \in \mathcal{H}_A \otimes \mathcal{H}_B$ is given by $|\psi\rangle_{AB} = \sum_i \lambda_i |i\rangle_A |i\rangle_B$, where $\lambda_i \geq 0$, $\sum_i \lambda_i^2 = 1$, and $|i\rangle_A$, $|i\rangle_B$ are orthonormal basis for $\mathcal{H}_A$, $\mathcal{H}_B$,

basis of $\mathcal{H}$. The Choi-Jamiołkowski matrix associated with the channel $\mathcal{E}$ is obtained by applying the channel $\mathcal{E}$ to the second half of $|\omega\rangle_{AB}$:

$$\sigma_{AB} = (\mathbb{1} \otimes \mathcal{E})(|\omega\rangle_{AB}). \tag{3.34}$$

The operator σ_{AB}, also known as dynamical matrix, is a $d^2 \times d^2$ operator acting on the Hilbert space $\mathcal{H}^{\otimes 2}$. The map $\mathcal{E}$ can be recovered by tracing out the system A from $\sigma_{AB} \cdot \rho^T \otimes \mathbb{1}$, i.e.,

$$\mathcal{E}(\rho) = \text{Tr}_A \left(\sigma_{AB} \cdot \rho^T \otimes \mathbb{1} \right). \tag{3.35}$$

Although we have supposed $\mathcal{E}$ as being a TPCP map, the Choi-Jamiołkowski isomorphism holds for any positive linear maps (in particular, TPCP maps). Moreover, the dynamical matrix σ_{AB} has special properties depending on the specificity of the corresponding linear map. For example, the set of completely positive maps is isomorphic to the set of all positive matrices acting on the corresponding composite space. In particular, if the operator $\mathcal{E}$ is a TPCP map, then the operator $\frac{1}{\text{Tr}(|\omega\rangle\langle\omega|_{AB})}\sigma_{AB}$ is a trace-one positive definite matrix. When the operator $\mathcal{E}$ is a unitary rotation (and hence a unitary operation), then the corresponding dynamical matrix is a density operator corresponding to a pure state in $\mathcal{H}^{\otimes 2}$ [3].

The Choi-Jamiołkowski isomorphism plays an important role in some problems of quantum information. When dealing with entanglement witness, for example, the isomorphism is employed to decide whether a given bipartite state ρ_{AB} is entangled or separable. Some results concerning the additivity conjecture of quantum information were demonstrated using this framework for representing quantum channels [11]. In the context of zero-error quantum communications, properties of the Choi-Jamiołkowski isomorphism are used to demonstrate the existence of quantum channels whose zero-error error capacity can be superactivated. More precisely, it was demonstrated the existence of two quantum channels, $\mathcal{E}_1$ and $\mathcal{E}_2$, for which no information can be perfectly transmitted when the channels are used individually, whereas perfect information transmission can be accomplished once the channels are used together, i.e., the channel $\mathcal{E}_1 \otimes \mathcal{E}_2$ has positive zero-error capacity.

Example 3.16 (The Choi-Jamiołkowski Isomorphism). Consider the quantum depolarizing channel discussed in Example 3.15:

$$\mathcal{E}(\rho) = (1-p)\rho + \frac{p}{3} \left(X\rho X + Y\rho Y + Z\rho Z \right).$$

---

respectively. The cardinality of $\{\lambda_i\}$, including multiplicity, is known as Schmidt rank or Schmidt number of $|\psi\rangle_{AB}$.

The Choi-Jamiołkowski matrix associated with the channel $\mathcal{E}$ is

$$\sigma_{AB} = (\mathbb{1}_d \otimes \mathcal{E})(|\omega\rangle_{AB})$$

$$= (1 - p)|\omega\rangle\langle\omega|_{AB} + \frac{p}{3}(\mathbb{1}_2 \otimes X \cdot |\omega\rangle\langle\omega|_{AB} \cdot \mathbb{1}_d \otimes X +$$

$$+ \mathbb{1}_d \otimes Y \cdot |\omega\rangle\langle\omega|_{AB} \cdot \mathbb{1}_d \otimes Y + \mathbb{1}_d \otimes Z \cdot |\omega\rangle\langle\omega|_{AB} \cdot \mathbb{1}_d \otimes Z),$$

where d is the dimension of the input/output Hilbert space $\mathcal{H}$, $|\omega\rangle_{AB} = \sum_{i=1}^{d} |i\rangle_A |i\rangle_B$ and $\{|1\rangle, \ldots, |d\rangle\}$ stands for the computational basis of $\mathcal{H}$. Equation (3.35) can be used to recover the Kraus operators of the channel or to calculate the output for a given input state.

To better illustrate this example, consider the depolarizing channel of two qubits with $p = 1/4$. Let $\{|0\rangle, |1\rangle\}$ be the computational basis of the 2-dimensional Hilbert space $\mathcal{H}$. The (up to a normalization factor) full Schmidt-rank state in $\mathcal{H}^{\otimes 2}$ is given by

$$|\omega\rangle_{AB} = |00\rangle + |11\rangle.$$

The dynamical matrix corresponding to the channel $\mathcal{E}$ is

$$\sigma_{AB} = \begin{bmatrix} \frac{5}{6} & 0 & 0 & \frac{6}{9} \\ 0 & \frac{1}{6} & 0 & 0 \\ 0 & 0 & \frac{1}{6} & 0 \\ \frac{6}{9} & 0 & 0 & \frac{5}{6} \end{bmatrix}.$$

Because $\mathcal{E}$ is a completely positive trace-preserving map, the matrix σ_{AB} is positive definite, i.e., all eigenvalues of σ_{AB} are positive. For instance, suppose we wish to calculate $\mathcal{E}(\rho)$, where $\rho = |\psi\rangle\langle\psi|$, with $|\psi\rangle = 0.6|0\rangle + 0.8|1\rangle$. The corresponding output state is given by

$$\mathcal{E}(\rho) = \mathrm{Tr}_A \left(\sigma_{AB} \cdot \rho^T \otimes \mathbb{1}_2 \right)$$

$$= \mathrm{Tr}_A \begin{pmatrix} 0.30 & 0.32 & 0.40 & \frac{32}{75} \\ 0 & 0.06 & 0 & 0.08 \\ 0.08 & 0 & \frac{8}{75} & 0 \\ 0.24 & 0.40 & 0.32 & \frac{8}{15} \end{pmatrix}$$

$$= \begin{pmatrix} \frac{61}{150} & 0.32 \\ 0.32 & \frac{89}{150} \end{pmatrix}.$$

The dynamical matrix (3.34) is known as *standard* Choi-Jamiołkowski matrix of $\mathcal{E}$ because it is obtained by setting $|\omega\rangle_{AB} = \sum_i |i\rangle_A |i\rangle_B$. In fact, the isomorphism holds for any bipartite, full Schmidt-rank state $|\omega\rangle_{AB} = \sum_i \lambda_i |\varphi_i\rangle_A |\chi_i\rangle_B$. Since A and B are identical Hilbert spaces, we can find a unitary basis change operator U

such that $U|\varphi_i\rangle = |\chi_i\rangle$. Therefore, the map (3.35) becomes

$$\mathcal{E}(\rho) = \mathrm{Tr}_A \left((U\sigma_A^{-1/2} \otimes \mathbb{1}) \cdot \sigma_{AB} \cdot (\sigma_A^{-1/2} U^\dagger \otimes \mathbb{1}) \cdot \rho^T \otimes \mathbb{1} \right), \tag{3.36}$$

where $\sigma_A = \mathrm{Tr}_B(\sigma_{AB})$. Finally, the non-standard (general) Choi-Jamiołkowski matrix σ_{AB} is related with the standard dynamic matrix $\tilde{\sigma}_{AB}$ by

$$\tilde{\sigma}_{AB} = (U\sigma_A^{-1/2} \otimes \mathbb{1}) \cdot \sigma_{AB} \cdot (\sigma_A^{-1/2} U^\dagger \otimes \mathbb{1}). \tag{3.37}$$

3.2.3 Accessible Information, Holevo Bound and $C_{1,1}$ Capacity of Quantum Channels

Consider a classical source emitting symbols $\mathcal{X} = 1,\ldots,n$ with probabilities $p_1,\ldots,p_n$. Suppose that symbols emitted by the source are used by Alice to prepare quantum states $\rho_1,\ldots,\rho_n$. After the preparation, Alice gives the quantum state to Bob, which is allowed to perform individual measurements aiming at inferring the symbol emitted by the source. Define X and Y as being the random variables representing the classical source and measurement outcomes, respectively. The *accessible information* is a measure of the maximum mutual information among these two random variables.

Definition 3.18 (Accessible Information). Let F be a quantum source emitting quantum states $\{\rho_1,\ldots,\rho_n\}$ with probabilities $p_1,\ldots,p_n$. Thus, this source is characterized by the ensemble $\{\rho_i, p_i\}$. Let X be a random variable that represents the source outputs. Let Y be a random variable describing measurement outcomes of the quantum states ρ_i. The accessible information between X and Y is given by

$$I_{acc}(F) = \max_{\{M_m\}} I(X;Y), \tag{3.38}$$

where the maximum is taken over all possible measurement schemes.

The accessible information is a measure of how well Bob can infer the state prepared by Alice. Unfortunately, no general method for calculating the accessible information is known; however, a variety of important bounds can be proved, the most important of which is the Holevo bound [24].

Definition 3.19 (Holevo Quantity [15]). Consider a quantum memoryless source F with ensemble $\{\rho_i, p_i\}$ of quantum states, i.e., the source F emits ρ_i with probability p_i. Define

$$\chi(F) = S(\rho) - \sum_i p_i S(\rho_i), \tag{3.39}$$

where S is the von Neumann entropy and $\rho = \sum_i p_i \rho_i$. The quantity $\chi(F)$ is known as *Holevo quantity*.

The Holevo bound states that

$$I_{acc}(F) \leq \chi(F). \tag{3.40}$$

It establishes an upper limit on how much information can be contained in a quantum system represented by a particular ensemble. Holevo bound and accessible information are discussed in more detail in Chap. 7.

Considering information measures and the characterization of quantum channels presented previously, we are ready to introduce the $C_{1,1}$ capacity of a quantum channel. This capacity can be understood as the maximum of the accessible information, at the channel output, over all ensembles of input states, once each output state is individually measured. [30].

Definition 3.20 ($C_{1,1}$ **Capacity [15, 30]**). Let $\mathcal{E}$ be a quantum channel as presented in Definition 3.17. The $C_{1,1}$ capacity of $\mathcal{E}$, denoted by $C_{1,1}(\mathcal{E})$, is defined as the maximum over all input ensembles of the accessible information of the corresponding output ensemble:

$$C_{1,1}(\mathcal{E}) = max_{\{\rho_x, p_x\}} I_{acc}(\{\mathcal{E}(\rho_x), p_x\}), \tag{3.41}$$

where I_{acc} is the accessible information of the ensemble $\{\mathcal{E}(\rho_x), p_x\}$.

The information transmission protocol of the $C_{1,1}$ capacity has three constraints:

1. Entangled states are not allowed between two or more uses of the channel $\mathcal{E}$. It explains the first "1" subscript in the $C_{1,1}$ capacity.
2. Joint quantum measurements involving several channel output are not allowed. This is the meaning of the second "1."
3. Adaptive measurements on the output are not allowed, i.e., Bob is not allowed to perform a partial measurement over the state, use such result to choose the next measurement, and return to complete the first measurement [22, 30].

Example 3.17 (Adapted from [30]). Consider that a quantum source can emit the states $|\psi_1\rangle$ and $|\psi_2\rangle$ according to a uniform probability distribution. Let $|\psi_1\rangle = |0\rangle$ and $|\psi_2\rangle = \cos\theta |0\rangle + \sin\theta |1\rangle$, where θ is a real parameter. The density operator ρ that represents this quantum source is

$$\rho = \frac{1}{2} \begin{bmatrix} 1 + \cos^2\theta & \cos\theta\sin\theta \\ \cos\theta\sin\theta & 1 - \cos^2\theta \end{bmatrix}.$$

The state ρ is sent over a quantum channel $\mathcal{E}$ which may introduce errors. Suppose that the quantum channel $\mathcal{E}$ is a binary symmetric channel with error probability

$$p = \frac{1}{2} - \frac{\sin\theta}{2}.$$

Fig. 3.10 A plot of the von
Neumann entropy of the
density matrix ρ and the
accessible information for
$\mathcal{E}(\rho)$, for $0 \leq \theta \leq \pi/2$

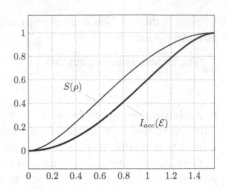

In this channel, the input $|\psi_1\rangle$ is replaced by $|\psi_2\rangle$ with probability p or left untouched with probability $1 - p$. The same holds for $|\psi_2\rangle$.

The von Neumann entropy of the input ρ is equal to $H(\frac{1}{2} - \frac{\cos\theta}{2})$. The accessible information is $1 - H(\frac{1}{2} - \frac{\sin\theta}{2})$. Both measures of information as a function of θ are illustrated in Fig. 3.10. It is possible to see that the von Neumann entropy is larger than the accessible information.

Considering this scenario, the value of θ which maximizes the output's accessible information is $\theta = \pi/2$. In this case,

$$C_{1,1}(\mathcal{E}) = 1 \text{ bit}, \tag{3.42}$$

which means that Bob can retrieve at most 1 bit per use of this quantum channel. The result is consistent with what we expected, since $p = 0$ (identity channel) for $\theta = \pi/2$.

3.2.4 Classical Capacity of a Quantum Channel

Quantum channels can be used in many different ways to transmit information between two parties. If we restrict ourselves to bits, the *classical capacity of a quantum channel* describes the amount of classical information that can be transmitted through the channel, a natural extension of the capacity definition from classical channels to the quantum world. This measure of information can be described by means of quantum mutual information [17].

Recall the problem of sending classical messages M randomly chosen from a set $\{M_1, M_2, \ldots, M_n\}$ by means of a quantum channel. Differently from the first assumption of the $C_{1,1}$ communication protocol, Alice is allowed to prepare codewords as tensor products of quantum states $\rho_1 \otimes \rho_2 \otimes \ldots$, where each of the states $\rho_1, \rho_2, \ldots$ is chosen from an ensemble $\{\rho_i, p_i\}$. Moreover, Bob is now allowed to use a POVM at the channel output. This means that he can wait for a number of states and measure all these states together (joint measurement), instead of measuring every qubit one by one (single measurement).

The notation $C_{1,\infty}(\mathcal{E})$ stands for the classical capacity of a quantum channel $\mathcal{E}$, in a scenario where Alice cannot use entangled states between two or more uses of the channel, whereas Bob is allowed to perform collective measurements at the channel output. The problem of finding $C_{1,\infty}$ was studied simultaneously and independently by Holevo [16] and by Schumacher and Westmoreland [27]. The following result is known as the *Holevo-Schumacher-Westmoreland (HSW) theorem.*

Theorem 3.4 (HSW Theorem [16, 27]). *The $C_{1,\infty}(\cdot)$ capacity of a quantum channel $\mathcal{E}$ is*

$$C_{1,\infty}(\mathcal{E}) \equiv \max_{\{\rho_i, p_i\}} \left[S\left(\mathcal{E}\left(\sum_i p_i \rho_i \right) \right) - \sum_i p_i S(\mathcal{E}(\rho_i)) \right], \qquad (3.43)$$

where the maximum is taken over all ensembles $\{p_i, \rho_i\}$ of input quantum states.

The proof of this theorem makes use of random coding and typical subspaces. A detailed demonstration can be found in Nielsen and Chuang [24, p. 555]. As with Shannon's channel coding theorem (vide Theorem 3.2), the capacity $C_{1,\infty}(\cdot)$ represents the maximum code rate for which the probability of transmission error can be made arbitrarily small, assuming sufficiently long message lengths [12].

Example 3.18 ($C_{1,\infty}$ Capacity of a Quantum Depolarizing Channel). Consider a 2-dimensional quantum depolarizing channel as presented in Example 3.15. Consider that Alice can send states from an ensemble $\{|\psi_j\rangle\langle\psi_j|, p_j\}$. Then, after passing through the depolarizing channel, we have

$$\mathcal{E}(|\psi_j\rangle\langle\psi_j|) = p|\psi_j\rangle\langle\psi_j| + (1-p)\frac{\mathbb{1}}{2}.$$

The quantum state $\mathcal{E}(|\psi_j\rangle\langle\psi_j|)$ has eigenvalues $(1+p)/2$. Therefore,

$$S(\mathcal{E}(|\psi_j\rangle\langle\psi_j|)) = H\left(\frac{1+p}{2}\right),$$

which does not depend on $|\psi_j\rangle$ at all. Hence, maximization (3.43) can be done by maximizing the entropy $S(\sum_j \mathcal{E}(|\psi_j\rangle\langle\psi_j|))$. Notice that if $\{|\psi_j\rangle\}$ is a set of orthonormal states, then $\sum_j \mathcal{E}(|\psi_j\rangle\langle\psi_j|) = p(\sum_j |\psi_j\rangle\langle\psi_j|) + (1-p)\mathbb{1}_2 = \mathbb{1}_2$, which maximizes $S(\sum_j \mathcal{E}(|\psi_j\rangle\langle\psi_j|))$. Therefore, the HSW capacity of the quantum depolarizing channel $\mathcal{E}$ is given by

$$C_{1,\infty}(\mathcal{E}) = 1 - H_2\left(\frac{1+p}{2}\right).$$

Figure 3.11 illustrates the plot of $C_{1,\infty}(\mathcal{E})$ versus the probability p. One can notice that the lower the probability of dephasing, the higher the channel capacity.

Fig. 3.11 A plot of the $C_{1,\infty}$ capacity for a quantum depolarizing channel versus the probability of non-dephasing

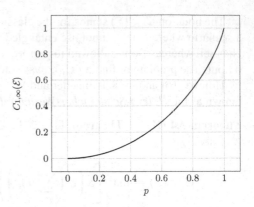

Example 3.19 (Classical Capacity of a Bit-Flip Quantum Channel [12]). Let $\mathcal{E}$ denote a bit-flip quantum channel with flip probability equal to p, as previously discussed in Example 3.14. Assuming orthogonal input symbols $\rho_1 = |0\rangle\langle 0|$ and $\rho_2 = |1\rangle\langle 1|$, the channel output is given by

$$\mathcal{E}\left(\sum_i p_i \rho_i\right) = \mathcal{E}(p_1 \rho_1 + (1 - p_1)\rho_2)$$

$$= \begin{pmatrix} p + p_1 - 2 \cdot p \cdot p_1 & 0 \\ 0 & 1 - [p + p_1 - 2 \cdot p \cdot p_1] \end{pmatrix}.$$

Notice that p stands for the bit-flip error probability, while p_1 stands for the probability associated with ρ_1 in the input ensemble. We are going to obtain the von Neumann entropies in order to calculate the channel capacity. First, we have

$$S\left[\mathcal{E}\left(\sum_i p_i \rho_i\right)\right] = -\{(p + p_1 - 2 \cdot p \cdot p_1)\log_2(p + p_1 - 2 \cdot p \cdot p_1)+$$

$$+ [1 - (p + p_1 - 2 \cdot p \cdot p_1)\log_2[1 - (p + p_1 - 2 \cdot p \cdot p_1)]]\}$$

$$= H(p + p_1 - 2 \cdot p \cdot p_1).$$

The von Neumann entropy of $\mathcal{E}(\rho_1)$ and $\mathcal{E}(\rho_2)$ is given by

$$S(\mathcal{E}(\rho_1)) = H(p),$$

$$S(\mathcal{E}(\rho_2)) = H(p).$$

Combining these results, we have that the $C_{1,\infty}$ capacity of the bit-flip quantum channel $\mathcal{E}$ is

$$C_{1,\infty}(\mathcal{E}) = \max_{\{p,p_1\}} H(p + p_1 - 2 \cdot p \cdot p_1) - H(p).$$

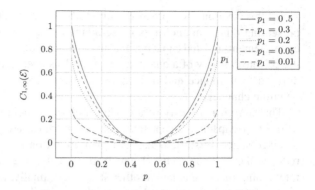

Fig. 3.12 A plot of the $C_{1,\infty}$ capacity for a bit-flip quantum channel considering different p_1 values

This maximization problem can be solved analytically by studying the partial derivatives, but we must recall that $\max S(\rho)$ is an upper bound for the capacity. Using the definition of the input density operator $\rho = p_1\rho_1 + (1 - p_1)\rho_2$, we get $S(\rho) = H(p_1)$. The maximum of $S(\rho)$ is attained considering a uniform distribution, i.e., when $p_1 = 1/2$. Therefore, for a bit-flip error probability p, the channel capacity is

$$C_{1,\infty}(\mathcal{E}) = H(p + 1/2 - 2p/2) - H(p)$$
$$= 1 - H(p). \tag{3.44}$$

Figure 3.12 shows the values of $C_{1,\infty}(\mathcal{E})$ where $\mathcal{E}$ is a bit-flip quantum channel. This capacity is shown as a function of the bit-flip probability p and for different values of p_1, the probability associated with one of the states in the input ensemble.

Interestingly, the channel capacity is seen to have two maxima, i.e., for $p = 0$ and $p = 1$. These two limiting cases correspond to the noiseless channel and the "deterministic" bit-flip channel, respectively. A deterministic bit-flipping is simply a change of code polarity, meaning that the classical codewords from Alice to Bob are exactly inverted or complemented, which entails no information degradation or error. The other limiting situation is obtained for $p = 1/2$, meaning that the qubit has a 50 % chance of being flipped and a 50 % chance of being conserved in its integrity. Bob's measurement amounts to a coin-flipping experiment, and all initial information is lost, which is the situation of the *useless channel* [12].

3.2.5 Other Capacities of Quantum Channels

A quantum communication channel can be used for several purposes: it can transmit classical information, private classical information, or quantum information. The channel can be used alone, with shared entanglement, or even together with other quantum channels. For each of these settings there is a capacity that quantifies a channel's potential for communication [33].

In this section, we proceed our study of quantum channels capacities by presenting other common types: the adaptive capacity of a quantum channel, the entanglement-assisted capacity, and also the quantum capacity. In Chap. 6, we cover the private capacity of a quantum channel. Besides, recall that most of this book is devoted to introduce and discuss the consequences of the zero-error capacity of quantum channels.

The *adaptive capacity* of a quantum channel, defined by Shor [31], is similar to the $C_{1,1}$ capacity, except that Bob can perform adaptive measurements on the received states. First, he makes a measurement on a given state that only partially reduces the quantum state. Then, Bob uses this measurement outcome to make intervening measurements on other states and, finally, he returns to make a further measurement on the reduced state of the original signal. The latter measurement may depend on the outcomes of intervening measurements [22].

In his paper, Shor illustrated the adaptive capacity using the lifted trine states:

$$T_0(\alpha) = \sqrt{1-\alpha} \, |000\rangle + \sqrt{\alpha} \, |001\rangle, \tag{3.45}$$

$$T_1(\alpha) = -\frac{1}{2}\sqrt{1-\alpha} \, |000\rangle + \frac{\sqrt{3}}{2}\sqrt{1-\alpha} \, |010\rangle + \sqrt{\alpha} \, |001\rangle, \tag{3.46}$$

$$T_2(\alpha) = -\frac{1}{2}\sqrt{1-\alpha} \, |000\rangle - \frac{\sqrt{3}}{2}\sqrt{1-\alpha} \, |010\rangle + \sqrt{\alpha} \, |001\rangle. \tag{3.47}$$

If the lifted trine states are used, then the adaptive capacity is strictly greater than the $C_{1,1}$ capacity, and less than the $C_{1,\infty}$ capacity, for $\alpha > 0$. Moreover, it was shown that for any ensemble of two pure states at the channel input, the adaptive capacity is equal to the $C_{1,1}$ capacity.

The adaptive capacity of a quantum channel is formalized below.

Definition 3.21 (Adaptive Capacity [31]). The adaptive capacity $C_{1,A}(\mathcal{E})$ of a quantum channel $\mathcal{E}$ is defined to be the supremum of the information rate over all encodings and all measurement strategies that use quantum operations local to the separate states and classical computation to coordinate them.

As previously introduced in Sect. 2.5, entanglement is a very interesting feature of quantum mechanics that is used as a physical resource by several protocols and applications in quantum information and computation. Regarding quantum channels, Bennett and his collaborators [4, 5] have demonstrated that shared entanglement can increase the classical capacity of noisy quantum channels [22].

The *entanglement-assisted capacity* of a noisy quantum channel is defined as the asymptotical classical information transmission rate that could be sent through a noisy quantum channel in a scenario where an arbitrary amount of entanglement is shared between Alice and Bob prior to the transmission.

Definition 3.22 (Entanglement-Assisted Capacity [5]). The entanglement-assisted capacity of a noisy quantum channel $\mathcal{E}$ is

$$C_E(\mathcal{E}) = \max_{\rho \in \mathcal{H}_{in}} S(\rho) + S(\mathcal{E}(\rho)) - S((\mathcal{E} \otimes \mathcal{I})(\Phi_\rho)), \tag{3.48}$$

where $\rho \in \mathcal{H}_{in}$ is a density matrix over the input states; Φ_ρ is a pure state belonging to the tensor product of state spaces $\mathcal{H}_{in} \otimes \mathcal{H}_R$ such that $\mathrm{Tr}_R[\Phi_\rho] = \rho$; $\mathcal{H}_{in}$ is the input state space and $\mathcal{H}_R$ is a space of reference. The third term of the right side of (3.48), $S((\mathcal{E} \otimes \mathcal{I})(\Phi_\rho))$, denotes the von Neumann entropy of the purification Φ_ρ of ρ over the reference system $\mathcal{H}_R$, half of which ($\mathcal{H}_{in}$) has been sent through the channel $\mathcal{E}$ while the other half ($\mathcal{H}_R$) has been sent through the identity channel (this corresponds to the portion of the entangled state that Bob holds at the start of the protocol) [22].

The quantity being maximized in (3.48) is the quantum mutual information. In order to transmit information using the protocol described above, Alice and Bob "consume" entanglement. In general, $S(\rho)$ qubits of entanglement (i.e., EPR pairs) per channel use are necessary to reach the entanglement-assisted capacity.

A remarkable thing about (3.48) is that we have an equality—the formula is *single-letter*. There is no need to take a limit over many channels uses, so this formula gives a complete characterization of the channel's capability for classical transmission given free access to entanglement [33]. It can also be concluded that shared entanglement does not change the additivity of quantum mutual information [17].

So far, we have presented different kinds of classical capacities for quantum channels. However, quantum channels extend the possibilities, and we can send classical information, entanglement-assisted classical information, private information, and, of course, *quantum information* [17].

To characterize this scenario, Alice has a quantum system whose state she would like to transmit coherently to Bob. To do so, Alice first encodes this quantum system into non-orthogonal superposed or entangled quantum states from a certain ensemble, sending the encoded state through a noisy quantum channel. Bob, in his turn, receives a quantum state which will be decoded. We are interested in answering how much quantum information can be reliably transmitted over a noisy quantum channel. The answer to this question is a quantity called *quantum capacity*, which also characterizes the fundamental limits of quantum error correction [33].

Before introducing the quantum capacity, we need to introduce the notion of *fidelity*, a measure that characterizes the quality of a quantum message transmission. The fidelity measures the probability of confusing two states if we are allowed to make only one measurement on one system, prepared in one of the two states [36].

Definition 3.23 (Fidelity). Let ρ and σ be two density operators. The fidelity F between ρ and σ is given by

$$F(\rho, \sigma) = \mathrm{Tr}\left(\sqrt{\sqrt{\rho}\,\sigma\,\sqrt{\rho}} \right). \tag{3.49}$$

Fidelity is a concept that comes from the inner product of two quantum states. Let $|\psi\rangle$ and $|\varphi\rangle$ be two states. The inner product $|\langle\varphi|\psi\rangle|^2$ gives the probability of finding the system in the state $|\varphi\rangle$ if it is known to be in the state $|\psi\rangle$, and vice versa. Hence this is a kind of measure of how similar the two states are or how much overlap there is between them. Suppose that $\rho = |\psi\rangle \langle\psi|$ and $\sigma = |\varphi\rangle \langle\varphi|$. Since they are pure states, $\rho^2 = \rho$ and $\sigma^2 = \sigma$, and hence, $\rho = \sqrt{\rho}$ and $\sigma = \sqrt{\sigma}$. Therefore, the fidelity of these two pure states is $F(\rho, \sigma) = |\langle\varphi|\psi\rangle|$ [21].

The fidelity is symmetric, i.e., $F(\rho, \sigma) = F(\sigma, \rho)$; it is invariant under unitary operations, that is, $F(U\rho U^\dagger, U\sigma U^\dagger) = F(\rho, \sigma)$. Because the fidelity represents a probability, $F(\rho, \sigma)$ is a number that ranges between 0 and 1, i.e., $0 \le F(\rho, \sigma) \le 1$.

Consider that Alice has a quantum source which emits states $\rho^{(1)}, \rho^{(2)}, \ldots,$ $\rho^{(n)}, \ldots$, where $\rho^{(n)} \in \mathcal{H}_{in}^{\otimes n}$ and $\mathcal{H}_{in}$ is the input Hilbert space. We define the entropy rate R of this source as

$$R = \limsup_{n \to \infty} \frac{S\left(\rho^{(n)}\right)}{n}, \tag{3.50}$$

where $S(\cdot)$ is the von Neumann entropy.

In order to send a state emitted by this source through a quantum channel $\mathcal{E}$, Alice needs to *encode* the state according to some encoding function $E_{nc} : B(\mathcal{H}_{in}^{\otimes n}) \to B(\mathcal{H}_{\mathcal{E}}^{\otimes m})$, where $\mathcal{H}_{in}$ is the input Hilbert space and $\mathcal{H}_{\mathcal{E}}$ is the channel Hilbert space. Upon receiving the quantum state at the channel output, Bob must *decode* it using a certain decoding function $D_{ec} : B(\mathcal{H}_{out}) \to B(\mathcal{H}_{in}^{\otimes n})$. We say that the pair (E_{nc}, D_{ec}) is a *coding scheme* for Alice's source into the channel $\mathcal{E}$, where E_{nc} and D_{ec} are trace preserving maps.

We say that a rate R is achievable over a quantum channel $\mathcal{E}$ if there is a quantum source with entropy R that may be sent reliably over the channel. We define the *quantum capacity* of the channel $\mathcal{E}$, denoted by $Q(\mathcal{E})$, as the supremum of all achievable rates over the channel $\mathcal{E}$.

Definition 3.24 (Quantum Capacity [20]). The quantum capacity of a quantum channel $\mathcal{E}$ is the largest number $Q(\mathcal{E})$ such that for any rate $R < Q(\mathcal{E})$, $\epsilon > 0$, and block sizes n and m, there exists an encoding procedure E_{nc} mapping n qubits into $m > n$ intermediate systems, $E_{nc}(\rho^{(n)})$, and feeds them to m independent instances of a quantum channel characterized by the superoperator $\mathcal{E}$; there should also be a decoding procedure D_{ec} mapping the m channel outputs to n qubits such that the original state $\rho^{(n)}$ can be recovered with a fidelity F at least $1 - \epsilon$, i.e.,

$$F(\rho^{(n)}, D_{ec}(\mathcal{E}^{\otimes m}(E_{nc}(\rho^{(m)})))) = 1 - \epsilon. \tag{3.51}$$

In order to introduce an important theorem regarding the quantum capacity, it is necessary to rethink how information is modified when it is transferred through a noisy quantum channel $\mathcal{E}_{AB}$ between Alice and Bob. We consider that the system of interest emitted by Alice's source, say ρ_A, interacts with the environment E. The environment starts in a pure state and the evolution is described by a quantum closed system, which covers the system of interest and the environment. The output of the quantum channel to Bob is the result of a partial trace over the state of the environment. In a mathematical notation, a noisy evolution through $\mathcal{E}$ has the form:

$$\mathcal{E}(\rho_A) = \text{Tr}_E\left[U\rho_A \otimes |0\rangle \langle 0|_E U^\dagger\right], \tag{3.52}$$

where $|0\rangle$ is some fixed pure state in which the environment starts and U is a unitary matrix from AE to BE [33].

We now introduce another quantity very important for quantum information theory, the *coherent information*. It is a property of a quantum state ρ_A and a quantum channel $\mathcal{E}$, which attempts to describe how much quantum information in the state will remain after the state goes through the channel. Coherent information is the quantum analogue of the classical mutual information, defined previously in Sect. 3.1.1.

Definition 3.25 (Coherent Information [26]). The coherent information between a state ρ and a quantum channel $\mathcal{E}$ is defined as

$$I(\rho\rangle\mathcal{E}) \equiv S(\mathcal{E}(\rho)) - S(\mathcal{E}, \rho), \tag{3.53}$$

where $S(\mathcal{E}(\rho))$ is the von Neumann entropy of the output state and $S(\mathcal{E}, \rho)$ is the entropy exchange between the state and the channel.

Remember that the Shannon capacity of a classical channel is calculated in terms of the mutual information. Analogously, the quantum channel capacity to carry quantum information is given in terms of coherent information. This result was stated and proved by Lloyd [19], Shor [29], and Devetak [13]. The *LSD theorem* states that

$$Q(\mathcal{E}) = \lim_{n \to \infty} \frac{1}{n} Q^{(1)}(\mathcal{E}^{\otimes n})$$

$$= \lim_{n \to \infty} \frac{1}{n} \max_{\{\rho_i, p_i\}} I(\rho_A\rangle\mathcal{E}(\rho_A))$$

$$= \lim_{n \to \infty} \frac{1}{n} \max_{\{\rho_i, p_i\}} (S(\rho_B) - S(\rho_E)), \tag{3.54}$$

where ρ_B is the channel output and ρ_E is the state of the environment [17]. The normalization factor $1/n$ indicates that the information transfer is measured per channel use. A detailed proof of this theorem can be found in Wilde [39, Sect. 23.n.5].

For a certain quantum channel $\mathcal{E}$, typically, $Q(\mathcal{E}) < C(\mathcal{E}) < C_E(\mathcal{E})$ [20]. Detailed examples of how to calculate the quantum capacity for amplitude damping and erasure quantum channels can be found in Wilde [39, Sect. 23.7].

3.3 Further Reading

In this chapter we introduced fundamental concepts of information theory by presenting it in two parts: the classical information theory and the quantum information theory. Regarding the classical part, we introduced measures of information, the notion of channel capacity, and two important theorems (noiseless coding and channel coding), both proposed by Shannon. Regarding the quantum part, we introduced some important information measures, the notion of quantum channels, and some different kinds of quantum channel capacities, such as the classical, entanglement-assisted, and quantum capacities.

There are many interesting topics regarding information theory that could not be covered here for matters of space and time. Regarding the classical part, some acclaimed surveys recall the development and the main results of this theory [6, 32, 37]. Also, many books with a more didactic approach have been published with exercises and examples, aiming at introducing this theory to the novice reader [9, 10, 14, 18, 23, 35].

Regarding quantum information theory, we refer the reader to very up-to-date books that comprise many concepts of this theory, discussing results, detailing proofs, and also presenting references to the seminal papers [12, 17, 20, 24, 36, 39]. In particular, for more details regarding the capacity of quantum channels, we recommend the reader to take a look at the following papers [2, 4, 19, 25, 26].

References

1. Arimoto S (1972) An algorithm for calculating the capacity of an arbitrary discrete memoryless channel. IEEE Trans Inf Theory 18:14–20
2. Barnum H, Nielsen MA, Schumacher B (1997) Information transmission through a noisy quantum channel. http://arxiv.org/abs/quant-ph/9702049. Accessed 25 Mar 2016
3. Bengtsson I, Zyczkowski K (2006) Geometry of quantum states. Cambridge University press, Cambridge
4. Bennett CH, DiVincenzo DP, Smolin JA, Wootters WK (1996) Mixed state entanglement and quantum error correction. Phys Rev A 54:3824
5. Bennett CH, Shor PW, Smolin JA, Thapliyal AV (2002) Entanglement-assisted capacity of a quantum channel and the reverse Shannon theorem. IEEE Trans Inf Theory 48:2637–2655
6. Berlekamp E (1974) Key papers in the development of coding theory. IEEE Press, New York
7. Blahut R (1972) Computation of channel capacity and rate distortion functions. IEEE Trans Inf Theory 18:460–473
8. Bruen AA, Forcinito MA (2005) Cryptography, information theory, and error-correction. Wiley-Interscience, New York
9. Cover TM, Thomas JA (2006) Elements of information theory. Wiley, New York
10. Csiszár I, Körner J (2011) Information theory? Coding theorems for discrete memoryless systems, 2nd edn. Cambridge University Press, Cambridge
11. Cubitt T, Harrow AW, Leung D, Montanaro A, Winter A (2008) Counterexamples to additivity of minimum output p-rényi entropy for p close to 0. Commun Math Phys 284(1):281–290. doi:10.1007/s00220-008-0625-z, http://dx.doi.org/10.1007/sn-0625-z
12. Desurvire E (2009) Classical and quantum information theory. Cambridge University Press, Cambridge
13. Devetak I (2005) The private classical capacity and quantum capacity of a quantum channel. IEEE Trans Inf Theory 51(1):44–55
14. Gray RM (2011) Entropy and information theory, 2nd edn. Springer, New York
15. Holevo AS (1973) Information theoretical aspects of quantum measurements. Probl Inf Transm 9(2):110–118

16. Holevo AS (1998) The capacity of the quantum channel with general signal states. IEEE Trans Inf Theory 4(1):269–273
17. Imre S, Gyongyosi L (2012) Advanced quantum communications: an engineering approach, 1st edn. Wiley, New York
18. Jones GA, Jones JM (2000) Information and coding theory, 1st edn. Springer, London
19. Lloyd S (1997) Capacity of the noisy quantum channel. Phys Rev A 55:1613–1622
20. Marinescu DC, Marinescu GM (2011) Classical and quantum information, 1st edn. Academic, New York
21. McMahon D (2008) Quantum computing explained, 1st edn. Wiley, New York
22. Medeiros RAC (2008) Zero-error capacity of quantum channels. Ph.D Thesis, Universidade Federal de Campina Grande – TELECOM Paris Tech
23. Moser SM, Chen PN (2012) A student's guide to coding and information theory, 1st edn. Cambridge University Press, Cambridge
24. Nielsen MA, Chuang IL (2010) Quantum computation and quantum information. Cambridge University Press, Cambridge
25. Schumacher B (1996) Sending quantum entanglement through noisy channels. http://arxiv.org/abs/quant-ph/9604023v1. Accessed 25 Mar 2016
26. Schumacher B, Nielsen MA (1996) Quantum data processing and error correction. http://arxiv.org/abs/quant-ph/9604022v1. Accessed 25 Mar 2016
27. Schumacher B, Westmoreland MD (1997) Sending classical information via noisy quantum channels. Phys Rev A 56. doi:10.1103/PhysRevA.56.131
28. Shannon CE (1948) A mathematical theory of communication. Bell Syst Tech J 27:623–656
29. Shor PW (2002) The quantum channel capacity and coherent information. In: MSRI Workshop on Quantum Computation, Berkeley, CA, pp 1–17. Available at: http://www.msri.org/realvideo/ln/msri/2002/quantumcrypto/shor/1/
30. Shor P (2003) Capacities of quantum channels and how to find them. http://arxiv.org/abs/quant-ph/0304102. Accessed 07 Mar 2016
31. Shor PW (2004) The adaptive classical capacity of a quantum channel, or information capacities of three symmetric pure states in three dimensions. IBM J Res & Dev 48(1):115–137
32. Slepian D (1973) Key papers in the development of information theory. IEEE, New York
33. Smith G (2010) Quantum channel capacities. http://arxiv.org/abs/1007.2855. Accessed 19 Mar 2016
34. Timpson CG (2013) Quantum information theory and the foundations of quantum mechanics, 1st edn. Oxford University Press, Oxford
35. Togneri R, deSilva CJS (2006) Fundamentals of information theory and coding design, 1st edn. Taylor & Francis, Boca Raton
36. Vedral V (2006) Introduction to quantum information science. Oxford University Press, Oxford
37. Verdu S, McLaughlin S (2000) Information theory: 50 years of discovery. IEEE, New York
38. von Neumann J (1955) Mathematical foundations of quantum mechanics. Princeton University Press, Princeton
39. Wilde MM (2013) Quantum information theory. Cambridge University Press, New York
40. Williams CP (2011) Explorations in quantum computing, 2nd edn. Springer, New York
41. Yanofsky NS, Mannucci MA (2008) Quantum computing for computer scientists. Cambridge University Press, Cambridge

Chapter 4
Classical Zero-Error Information Theory

Information theory was introduced by Claude E. Shannon in 1948 [10]. In his paper, Shannon defined a number C representing the capacity of a communication channel for transmitting information reliably. He proved the existence of codes that allow reliable transmission between two parties, provided that the communication rate is less than the channel capacity. A randomly generated code with large block size has a high probability to be a good code. By reliable transmission we mean that the error probability can be made as close to zero as possible, but not actually zero. Most of the information theory issues, including channel capacity, are based on probability theory and statistics. This asymptotic capacity is hereafter denoted as the *ordinary capacity*.

In 1956, 8 years after his first paper introducing information theory, Shannon demonstrated how discrete memoryless channels (DMCs) could be used to transmit information in a scenario where *no errors* are permitted, instead of allowing an asymptotically small probability of error. The so-called *zero-error capacity* was defined as the least upper bound of rates at which information can be transmitted through a DMC with a probability of error equal to zero [11].

Most of the real communication systems do not require a zero probability of error when transmitting and receiving information. Although, there are some situations in which it would be interesting to consider a scenario where no transmission errors are allowed and ask for the maximum rate at which information can be transmitted [6]:

- Applications where no errors can be tolerated;
- In some models, only a small number of channel uses or a few source instances are available. Therefore, we cannot appeal to results ensuring that the error probability decreases as the number of uses or instances increases;
- Zero-error information theory can be used to study the communication complexity of error-free protocols and functions;
- Functionals and methods originally used in zero-error information theory are often applied in mathematics and computer science.

© Springer International Publishing Switzerland 2016
E.B. Guedes et al., *Quantum Zero-Error Information Theory*,
DOI 10.1007/978-3-319-42794-2_4

In his seminal paper, Shannon gave a graph theoretic approach to the zero-error capacity. By associating a DMC with a graph, Shannon introduced a new quantity in graph theory, the *Shannon capacity of a graph* [1, 4, 5]. Differently from the ordinary capacity, finding the zero-error capacity of a DMC (or a graph) is a combinatorial problem. Because of its restrictive nature—a vanishing probability of error is required—the zero-error information theory is frequently unknown to many information theorists. Nevertheless, its methods play an important role in areas like combinatorics and graph theory.

This chapter aims to introduce the main concepts of the classical zero-error information theory. We begin by defining the zero-error capacity of a discrete memoryless channel in Sect. 4.1. Section 4.2 shows how the problem of finding the zero-error capacity can be stated in terms of a graph quantity. The Lovász theta function and some of its properties are depicted in Sect. 4.3 and, lastly, in Sect. 4.4, the zero-error capacity of the sum and product of channels is discussed.

4.1 The Zero-Error Capacity

In communication theory, we say that Alice communicates with Bob when the physical acts of Alice have induced a desired physical state in Bob. As this transfer of information is a physical process, it is subject to the uncontrollable ambient noise and imperfections of the physical signaling process itself. The communication is successful if the receiver Bob and the transmitter Alice agree on what was sent.

A common classical communication system was shown in Fig. 3.3. The encoder maps source symbols from a finite alphabet into some sequence of channel symbols, afterwards called codeword, which is sent through the channel. The channel produces an output sequence which is random but has a probability distribution that depends on the input sequence. From the output sequence, we attempt to recover the transmitted message. Two input sequences are said to be *confusable* when they induce the same output sequence. Shannon showed that we can choose a "non-confusable" subset of input sequences in a manner that with high probability, there is only one highly likely input that could have caused the particular output. Essentially, this means that we can reconstruct input sequences at the output with negligible probability of error. As already described in Chap. 3, the maximum rate at which this can be done is called the ordinary capacity of the channel C.

The channel coding theorem asserts that even the best coding scheme attaining the ordinary capacity C allows for an asymptotically small but non-vanishing probability of error. From now, we will be interested in the case where no transmission errors are permitted.

Consider a classical discrete memoryless channel $(\mathcal{X}, \mathcal{Y}, P)$. Symbols in the input and output alphabets will be hereafter called input and output symbols, respectively. Shannon [11] defined an error-free code as follows:

Definition 4.1 ((M, n) Error-Free Code). An (M, n) error-free code for the DMC $(\mathcal{X}, \mathcal{Y}, P)$ in Fig. 3.3 is composed of the following:

1. A set of indexes $\{1, \ldots, M\}$, where each index is associated with a classical message.
2. An encoding function

$$X^n : \{1, \ldots, M\} \to \mathcal{X}^n,$$

yielding codewords $\mathbf{x}^1 = X^n(1), \ldots, \mathbf{x}^M = X^n(M)$. The set of all codewords is called a codebook.
3. A decoding function

$$g : \mathcal{Y}^n \to \{1, \ldots, M\},$$

which deterministically assigns a guess to each possible received codeword with the following property:

$$\Pr\left(g(Y^n) \neq i | X^n = X^n(i)\right) = 0 \; \forall \; i \in \{1, \ldots, M\}. \tag{4.1}$$

In the zero-error context, we are particularly interested in symbols that can be fully distinguished at the channel output. They are called non-adjacent symbols.

Definition 4.2 (Adjacency). Consider a DMC $(\mathcal{X}, \mathcal{Y}, P)$. Two input symbols $x_i, x_j \in \mathcal{X}$ are said to be adjacent (or indistinguishable) if there exists an output symbol in $\mathcal{Y}$ that can be caused by either of these two, i.e., there is an $y \in \mathcal{Y}$ such that both $p(y|x_i)$ and $p(y|x_j)$ do not vanish. Otherwise, they are said to be non-adjacent (or distinguishable).

Consider the sequence $\mathbf{x} = x_1 x_2 \ldots x_n$ being transmitted through a DMC. The output sequence $\mathbf{y} = y_1 y_2 \ldots y_n$ is received with probability

$$p^n(\mathbf{y}|\mathbf{x}) = \prod_{i=1}^{n} p(y_i|x_i). \tag{4.2}$$

If two sequences $\mathbf{x}'$ and $\mathbf{x}''$ can both result in the sequence $\mathbf{y}$ with positive probability, then no decoder can decide with zero probability of error which of the two sequences has been transmitted by the sender. Such sequences will be called *indistinguishable* or adjacent at the receiving end of the DMC. In fact, if all input symbols in $\mathcal{X}$ are adjacent to each other, any code with more than one codeword has a probability of error great than zero. This is equivalent to say that $\mathbf{x}'$ and $\mathbf{x}''$ are distinguishable if and only if there exists at least one i, $1 \leq i \leq n$, such that x_i' and x_i'' are non-adjacent, as illustrated in Fig. 4.1.

It is useful to think of probability distributions $p(y|x)$ and $p^n(\cdot|\mathbf{x})$ as vectors of dimension $|\mathcal{X}|$ and $|\mathcal{X}|^n$, respectively. Using this approach, we can rewrite the statement given earlier by saying that two sequences $\mathbf{x}', \mathbf{x}'' \in \mathcal{X}^n$ are distinguishable at the receiving end of the DMC channel if and only if the corresponding vectors $p^n(\cdot|\mathbf{x}')$ and $p^n(\cdot|\mathbf{x}'')$ are orthogonal.

$$
\begin{aligned}
\mathbf{x}' &= x_1' x_2' \ldots \boxed{x_i'} \ldots x_{n-1}' x_n' \\
\mathbf{x}'' &= x_1'' x_2'' \ldots \boxed{x_i''} \ldots x_{n-1}'' x_n''
\end{aligned}
$$

Fig. 4.1 Two distinguishable sequences $\mathbf{x}'$ and $\mathbf{x}''$. For at least one i, $1 \le i \le n$, the input symbols x_i' and x_i'' are non-adjacent

Definition 4.3 (Zero-Error Capacity). Define $N(n)$ as the maximum cardinality of a set of mutually orthogonal vectors among the $p^n(\cdot|\mathbf{x})$, $\mathbf{x} \in \mathcal{X}^n$. The zero-error capacity of the channel $(\mathcal{X}, \mathcal{Y}, P)$ is given by

$$
C_0 = \limsup_{n \to \infty} \frac{1}{n} \log N(n). \tag{4.3}
$$

Intuitively, C_0 is the bit-per-symbol error-free information transmission rate capability of the channel.

The number $N(n)$ in (4.3) is super multiplicative, i.e.,

$$
N(n + m) \ge N(n) \cdot N(m). \tag{4.4}
$$

To verify this, let $\mathbf{x}'$ and $\mathbf{x}''$ be sequences of length n and m, respectively. Then, there exist at least $N(n) \cdot N(m)$ non-adjacent sequences of length $n + m$, obtained by concatenating sequences of length n with sequences of length m. Hence, we can use the Fekete's lemma [12, p. 85] to demonstrate that the superior limit (4.3) is a true limit and actually coincides with the supremum of numbers $\frac{1}{n} \log N(n)$.

Shannon pointed out that the zero-error capacity of a DMC $(\mathcal{X}, \mathcal{Y}, P)$ depends only on which symbols in $\mathcal{X}$ are adjacent to each other. This is a major difference between the error-free capacity and the ordinary capacity, since in the latter the capacity depends on the choice of the probability distribution of the input symbols $\mathcal{X}$. It is easy to demonstrate that a DMC has a non-vanishing error-free capacity if and only if there exist at least two non-adjacent symbols in $\mathcal{X}$. Figure 4.2 shows some discrete memoryless channels. For the binary symmetric channel with $0 < p < 1$, the two input symbols are adjacent yielding $C_0 = 0$. Both channels in Fig. 4.2b, c have at most two pairs of non-adjacent symbols. For example, if we consider codewords of length one, we can perform error-free communication by choosing to send only symbols $\{0, 2\}$ or $\{1, 3\}$ of the channel in Fig. 4.2b. In this case, the rate of the code is $\log 2 = 1$ bit per channel use.

One might ask whether we can increase the transmission rate by varying the code length or whether $C_0 = \log N(1)$. It turns out that we can. Consider the sequences $\{00, 12, 24, 31, 43\}$ of length 2 for the G_5 DMC of Fig. 4.2c. Clearly, these sequences are pairwise distinguishable at the channel output and hence are codewords of an error-free code of length two. The ordinary capacity of G_5 can be easily calculated $(C = \log 5/2)$. Therefore, the zero-error capacity of G_5 is lower

Fig. 4.2 Some discrete memoryless channels. Since we are interested in adjacency relations, we omit the transition probabilities. (**a**) The binary symmetric channel. (**b**) A discrete channel with quaternary alphabeth. (**c**) The pengaton channel G_5

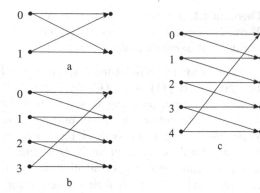

and upper bounded by

$$\frac{1}{2}\log 5 \le C_0(G_5) \le \log 5/2. \tag{4.5}$$

These bounds were given by Shannon in 1956, and the problem of finding the capacity $C_0(G_5)$ remained open during 20 years until Lovász [7] gave a brilliant solution. He showed that the Shannon's lower bound was tight

$$C_0(G_5) = \frac{1}{2}\log 5.$$

We demonstrate such result in Sect. 4.3, where we introduce the Lovász theta function.

As we can see, the calculation of the zero-error capacity is a very difficult problem even for simple channels. Although some methods we discuss in the next sections enable the computation of the zero-error capacity of particular classes of discrete memoryless channels, the general problem remains wide open.

4.1.1 The Adjacency-Reducing Mapping

The calculation of the zero-error capacity of simple channels can be done using the notion of *adjacency-reducing mapping*. This means a mapping $f : \mathcal{X} \rightarrow \mathcal{X}$ with the property that if x_i and x_j are not adjacent in the channel, then $f(x_i)$ and $f(x_j)$ are not adjacent. Given an (M, n) error-free code for the channel, we can obtain an equivalent error-free code by applying the adjacency-reducing mapping to each codeword, since f never produces new adjacencies. Suppose that for a given DMC the mapping f takes all symbols in $\mathcal{X}$ onto a subset $\mathcal{X}' \subset \mathcal{X}$ of the symbols no two of which are adjacent. Clearly, there are at least $|\mathcal{X}'|^n$ n-length distinguishable sequences for this channel. However, any error-free code of length n has at most $|\mathcal{X}'|^n$ sequences, given that the application of f on this code leads to a new error-free code whose alphabet contains only $|\mathcal{X}'|$ symbols. These observations imply the following theorem:

Theorem 4.1. *Let $(\mathcal{X}, \mathcal{Y}, P)$ be a discrete memoryless channel. If all symbols in $\mathcal{X}$ can be mapped by an adjacency-reducing mapping f into a subset $\mathcal{X}' \subset \mathcal{X}$ of non-adjacent symbols, then $C_0 = \log |\mathcal{X}'|$.*

As an example, consider the DMC illustrated in Fig. 4.2b. Let f be a mapping with $f(0) = 0, f(1) = 0, f(2) = 2$ and $f(3) = 2$. It is easy to see that f is an adjacency-reducing mapping satisfying the condition of Theorem 4.1, where $\mathcal{X}' = \{0, 2\}$. Therefore, the zero-error capacity of the channel is $C_0 = \log |\mathcal{X}'| = 1$ bit per channel use. It is easy to see that we cannot construct an adjacency-reducing mapping f for the G_5. Except for the G_5 channel, it is possible and easy to construct adjacency-reducing maps for all discrete memoryless channels up to five input symbols—and so calculate their zero-error capacities. All DMCs with six input symbols were analyzed and their zero-error capacity computed, except for four channels whose capacity can be given in terms of $C_0(G_5)$.

In the next section, we show how a graph (and its complement) can be associated with a discrete memoryless channel. Theorem 4.1 is restated in a graph-based language.

4.2   Relation with Graph Theory

The problem of computing the zero-error capacity of discrete memoryless channels can be reformulated in terms of graph theory. Given a DMC $(\mathcal{X}, \mathcal{Y}, P)$, we can construct a characteristic graph G as follows. Take as many vertices as the number of input symbols in $\mathcal{X}$ and connect two vertices with an edge if the corresponding input symbols in $\mathcal{X}$ are distinguishable. Using such approach, the vertex set of G is $V(G) = \mathcal{X}$ and its set of edges $E(G)$ is composed of pairs of orthogonal rows in P. The characteristic graph of channels in Fig. 4.2 is shown in Fig. 4.3.

In graph theory, the order of a graph is the cardinality of its vertex set. A *clique* is defined as any complete subgraph of G, and the *clique number* [2] of a graph G, denoted by $\omega(G)$, stands for the maximal order of a clique in G. It is easy to see that

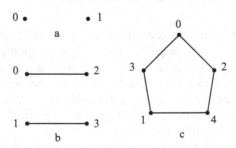

Fig. 4.3 Characteristic graphs G of discrete memoryless channels in Fig. 4.2. The vertex set of G is the set of input symbols $\mathcal{X}$ and its set of edges corresponds to all pairs of distinguishable symbols in $\mathcal{X}$. (**a**) Characteristic graph of the binary symmetric channel. (**b**) Characteristic graph of the discrete channel with quaternary alphabeth. (**c**) Characteristic graph of the pengaton channel G_5

the maximum number of non-adjacent symbols in G is $\omega(G)$, namely $N(1) = \omega(G)$. For example, the pentagon graph G_5 of Fig. 4.3c has the clique number $\omega(G_5) = 2$. Note that the vertex set of any clique corresponds to a set of distinguishable symbols in the channel.

Define the n-product G^n of the graph G as a graph for which $V(G^n) = \mathcal{X}^n$ and $\{\mathbf{x}', \mathbf{x}''\} \in E(G^n)$ if for at least one $1 \leq i \leq n$, the i-th coordinates of $\mathbf{x}'$ and $\mathbf{x}''$ satisfy $\{x_i', x_i''\} \in E(G)$. Such product of graphs, often called *Shannon's product*, has the following meaning: the vertex set of G^n is composed of all n-length sequences, and we connect the vertices $\mathbf{x}'$ and $\mathbf{x}''$ if the corresponding sequences are distinguishable, as illustrated in Fig. 4.1.

It is clear that the number of distinguishable sequences of length n is the clique number of G^n, i.e, $N(n) = \omega(G^n)$. Moreover, the sequences in the vertex set of the corresponding complete subgraph define an n-length error-free code for the channel. Therefore, the zero-error capacity of the DMC $(\mathcal{X}, \mathcal{Y}, P)$ can be rewritten as

$$C_0 = \sup_n \frac{1}{n} \log \omega(G^n). \tag{4.6}$$

In graph theory, the value (4.6) refers to the Shannon capacity of the graph G, and is denoted by $C_0(G)$.

The chromatic number of a graph G, denoted by $\chi(G)$, is the smallest number of colors necessary to color the vertices of a graph so that no two adjacent vertices have the same color. More formally, $\chi(G)$ is the smallest cardinality of a set K for which there exists a function $f : V(G) \to K$ with the property that adjacent vertices are mapped into different elements of K. Let $(\mathcal{X}, \mathcal{Y}, P)$ be a DMC for which the clique and the chromatic numbers of the characteristic graph G are the same, $\omega(G) = \chi(G)$. For any coloration of G, if we define the set $\mathcal{X}'$ in Theorem 4.1 as being the vertex set of the maximal clique in G, then we can always construct an adjacency-reducing mapping f fulfilling the requirement of the theorem: all symbols whose vertices share a given color are mapped into the corresponding symbol in $\mathcal{X}'$ that own such color. Because different colors are associated with non-adjacent symbols, such mapping ensures that any two non-adjacent symbols in $\mathcal{X}$ will be mapped into non-adjacent ones in $\mathcal{X}'$. Moreover, because symbols in $\mathcal{X}'$ correspond to the vertex set of the maximal clique, they are mutually distinguishable. Therefore, Theorem 4.1 can be entirely reformulated.

Theorem 4.1'. *Let $(\mathcal{X}, \mathcal{Y}, P)$ be a discrete memoryless channel and G the corresponding characteristic graph. If $\omega(G) = \chi(G)$, then $C_0 = \chi(G)$.*

The best known graphs for which $\omega(G) = \chi(G)$ are the so-called *perfect graphs* [2]. A perfect graph is a graph G such that for every induced subgraph of G, the chromatic number equals the clique number. The class of perfect graphs includes bipartite graphs, interval graphs, and wheel graphs with an odd number of vertices. The smallest vertex set on which a graph exists with $\omega(G) \neq \chi(G)$ has five vertices, and corresponds to the pentagon graph G_5 already discussed in the previous section.

Although $\omega(G) = \chi(G)$ is a sufficient condition for $\omega(G^n) = [\omega(G)]^n$, Lovász showed that it is not a necessary condition [7]. An example is the complement of the Petersen graph, which is isomorphic with the Kneser graph $KG_{5,2}$. However, it is unknown whether the equality $\log \omega(G') = C_0(G')$ for every induced subgraph $G' \subseteq G$ implies that G is perfect.

Originally, Shannon used a different but equivalent approach for relating the zero-error capacity with elements of graph theory. For a given DMC $(\mathcal{X}, \mathcal{Y}, P)$, we can associate an adjacency matrix $A = \begin{bmatrix} a_{ij} \end{bmatrix}$ as follows:

$$a_{ij} = \begin{cases} 1 & \text{if } x_i \text{ is adjacent to } x_j \text{ or if } i = j \\ 0 & \text{otherwise,} \end{cases} \tag{4.7}$$

where $x_i, x_j \in \mathcal{X}$. If two channels give rise to the same adjacency matrix, then it is obvious that an error-free code for one will be an error-free code for the other and, hence, the zero-error capacity C_0 for one will also apply to the other [11]. Such approach considers the adjacency structure of the adjacency matrix to construct a linear graph, called adjacency graph, which is the complementary of the characteristic graph. Therefore, both graphs have the same vertex set $\mathcal{X}$ and two vertices in the adjacency graph are connected by an edge if and only if they are not connected in the characteristic graph. Equivalently, an edge connects two vertices in the adjacency graph if and only if the corresponding input symbols in $\mathcal{X}$ are adjacent. In this case, we say that two vertices in the adjacent graph are independent if the corresponding symbols are non-adjacent in the channel. Clearly, there are $N(1)$ independent vertices in G. Figure 4.4 shows the adjacency graphs of the discrete memoryless channels of Fig. 4.2.

Shannon [11] proved the following bounds on the zero-error capacity:

Theorem 4.2. *Let $(\mathcal{X}, \mathcal{Y}, P)$ be a DMC. The error-free capacity is bounded by the inequalities:*

$$- \log \min_{p(x_i)} \sum_{ij} a_{ij} p(x_i) p(x_j) \le C_0 \le \min_{p(y|x)} C, \tag{4.8}$$

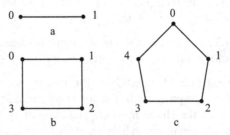

Fig. 4.4 Adjacency graphs of discrete memoryless channels corresponding to the channels of Fig. 4.2. These graphs are constructed by taking as many vertices as the number of symbols in $\mathcal{X}$, and connecting two vertices if the corresponding symbols are adjacent in the channel. (**a**) Adjacency graph of the binary symmetric channel. (**b**) Adjacency graph of the discrete channel with quaternary alphabeth. (**c**) Adjacency graph of the pengaton channel G_5

where C is the ordinary capacity of any discrete memoryless channel with stochastic matrix $P = [p(y|x)]$ *giving rise to the adjacency matrix* $A = [a_{ij}]$; $p(x_i)$ *stands for the input probability distribution.*

The proof of the theorem can be found in [11]. Although the upper bound is fairly obvious, it has an interesting formulation in graph theory [8] according to which

$$C_0 \leq \log \chi^*(G), \tag{4.9}$$

where $\chi^*(G)$ is the fractional chromatic number of the adjacency graph G, a well-studied concept in polyhedral combinatorics [9] defined as follows. We assign nonnegative weights $p(x_i)$ to the vertices $\mathcal{X}$ of G such that

$$\sum_{x_i \in C} p(x_i) \leq 1$$

for every complete subgraph C in G. This assignment is called a fractional coloring. The fractional chromatic number is the maximum of $\sum_{x_i \in \mathcal{X}} p(x_i)$, where the maximum is taken over all fractional colorings of G. Actually, the fractional chromatic number is the solution of the real-valued relaxation of the integer programming problem that defines the chromatic number of G [6].

Suppose that a DMC $(\mathcal{X}, \mathcal{Y}, P)$ gives rise to an adjacency graph G such that G can be covered by $N(1)$ cliques. By this we mean that there are $N(1)$ cliques in G, namely $C_1, \ldots, C_{N(1)}$, in a way that their vertex sets, $V(C_1), \ldots, V(C_{N(1)})$, form a partition of $V(G)$. Theorem 4.1 can be rewritten as [7]:

Theorem 4.1". *Let G be the adjacency graph of a discrete memoryless channel* $(\mathcal{X}, \mathcal{Y}, P)$. *If G can be covered by* $N(1)$ *cliques, then* $C_0 = \log N(1)$.

Figure 4.5 illustrates Theorem 4.1". The maximum number of independent vertices in the adjacency graph of Fig. 4.5a is $N(1) = 2$, e.g., 0 and 3. An adjacency-reducing mapping f for the corresponding DMC takes $f(0) = f(1) = f(2) = 0$ and $f(3) = f(4) = 3$. This mapping can be readily obtained by associating 0 and 3 with vertices of the order-2 and order-3 cliques, respectively. The cube graph has $N(1) = 4$, and can be covered by four clique of order 2 as illustrated in Fig. 4.5b. Therefore, the zero-error capacity of the equivalent DMC is $C_0 = \log 4 = 2$ bits per channel use.

4.3 Lovász ϑ Function

The redefinition of the zero-error capacity in terms of graph has yielded interesting constructions in combinatorics and graph theory. An example of such constructions is the *Lovász theta function* ϑ. The functional ϑ has many applications in computer science and combinatorics. Particularly, the ϑ function is a polynomially

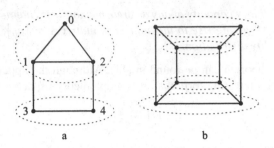

a   b

Fig. 4.5 Graphs that can be covered by a number of cliques. (**a**) An adjacency graph with two independent vertices. This graph can be covered by two cliques and therefore there is an adjacency reducing map satisfying the requirement of Theorem 4.1. (**b**) The cube graph can be covered by four cliques of order two

computable functional sandwiched in between two NP-complete problems in graph theory: the clique and the chromatic numbers of a graph [3]. The very nice formulation we present in this section was used to compute the zero-error capacity of the pentagon graph (4.5), initially lower and upper bounded by Shannon.

Given a DMC $(\mathcal{X}, \mathcal{Y}, P)$ and the corresponding adjacency graph G with vertex set $\mathcal{X}$, an orthonormal representation of G is a set of $|\mathcal{X}|$ vectors $\mathbf{v}_{x_i}$ in a Euclidean space, such that if $x_i, x_j \in \mathcal{X}$ are non-adjacent, then $\mathbf{v}_{x_i}$ and $\mathbf{v}_{x_j}$ are orthogonal. The *value* of an orthonormal representation is defined as

$$\min_{\mathbf{c}} \max_{x_i \in \mathcal{X}} \frac{1}{(\mathbf{c}^T \mathbf{v}_{x_i})^2},$$

where the minimum is taken over all unitary vectors $\mathbf{c}$. The vector $\mathbf{c}$ yielding the minimum is called the handle of the representation. The Lovász $\vartheta(G)$ function of a graph is defined as the minimum value over all representations of G, and a representation is called optimal if it attains this minimum value. Lovász proved the following result [7]:

Theorem 4.3. *The zero-error capacity of a DMC $(\mathcal{X}, \mathcal{Y}, P)$ is upper bounded by the logarithm of the ϑ function of its adjacency graph G:*

$$C_0 \leq \log \vartheta(G). \tag{4.10}$$

Proof. First, we should note that if G and H are two graphs, and GH is their product as defined in Sect. 4.2, then $\vartheta(GH) \leq \vartheta(G)\vartheta(H)$. Let $\{\mathbf{v}_{x_i'}\}$ and $\{\mathbf{u}_{x_j''}\}$ be optimal orthonormal representations of G and H with handles $\mathbf{c}$ and $\mathbf{d}$, respectively. It is easy to see that $\{\mathbf{v}_{x_i'} \otimes \mathbf{u}_{x_j''}\}$ is an orthonormal representation of GH and $\mathbf{c} \otimes \mathbf{d}$ is a unitary vector. Therefore,

$$\vartheta(GH) \leq \max_{x_i', x_j''} \frac{1}{\left((\mathbf{c} \otimes \mathbf{d})^T (\mathbf{v}_{x_i'} \otimes \mathbf{u}_{x_j''})\right)^2}$$

$$= \max_{x_i'} \frac{1}{(\mathbf{c}^T \mathbf{v}_{x_i})^2} \max_{x_j''} \frac{1}{(\mathbf{d}^T \mathbf{u}_{x_j''})^2}$$

$$= \vartheta(G)\vartheta(H).$$

By definition, if G is an adjacency graph and $\{\mathbf{v}_{x_i}\}$ is an optimum representation with handle $\mathbf{c}$, then there are $N(1)$ vectors $\{\mathbf{v}_{x_1}, \ldots, \mathbf{v}_{x_{N(1)}}\}$ pairwise orthogonal in this representation, where $N(1)$ is the maximum number of independent vertices in G. Hence,

$$1 = \|\mathbf{c}\|^2 \geq \sum_{i=1}^{N(1)} (\mathbf{c}^T \mathbf{v}_{x_i})^2 \geq \frac{N(1)}{\vartheta(G)}. \tag{4.11}$$

Equation (4.4) implies $N(1)^n \leq N(n)$. Finally,

$$C_0 = \sup_n \frac{1}{n} \log N(n) \leq \sup_n \frac{1}{n} \log \vartheta(G^n) \leq \sup_n \frac{1}{n} \log \vartheta(G)^n = \log \vartheta(G).$$

Theorem 4.3 allows of the calculation of the zero-error capacity of the pentagon graph. Remember that Shannon was only able to give lower and upper bounds for the capacity, $\frac{1}{2} \log 5 \leq C_0(G_5) \leq \log \frac{5}{2}$. First, construct an orthonormal representation for the pentagon G_5 of Fig. 4.4c as follows. Consider an umbrella whose handle and five ribs have unitary length. Let $\mathbf{v}_0, \ldots, \mathbf{v}_4$ be the ribs and $\mathbf{c}$ the handle, as vectors oriented away from their common point. Open the umbrella to the point where the maximum angle between the ribs is $\pi/2$. Note that the angle between two consecutive ribs must be the same, and that the angle between alternate ribs must be $\pi/2$. It is clear that $\{\mathbf{v}_0, \ldots, \mathbf{v}_4\}$ forms an orthonormal representation of G_5. Figure 4.6 illustrates this scenario, at which we plot the handle $\mathbf{c}$ and the two orthogonal vectors $\mathbf{v}_1$ and $\mathbf{v}_3$. The extremities of the six vectors are points on a

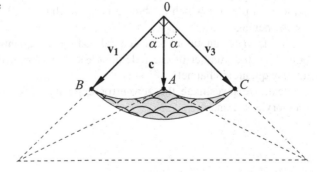

Fig. 4.6 A spherical triangle delimited by the vectors $\mathbf{v}_1$, $\mathbf{v}_3$ and the handle $\mathbf{c}$. In a plane normal to the handle, the angle between two consecutive projections $\mathbf{v}'_i, \mathbf{v}'_{i+1 \mod 5}$ of the vectors $\mathbf{v}_i$ is $2\pi/5$. The spherical angle $\angle A$ is the angle between the vectors $\mathbf{v}'_1$ and $\mathbf{v}'_3$, i.e., $\angle A = 4\pi/5$

unitary three-dimensional sphere centered in 0, and the points defined by the handle and any two alternated vectors delimit a spherical triangle identical to the triangle ABC of Fig. 4.6. We are interested in the value of the representation, i.e.,

$$\min_{\mathbf{c}} \max_{0 \le i \le 4} \frac{1}{(\mathbf{c}^T \mathbf{v}_i)^2}.$$

Note that $\mathbf{c}^T \mathbf{v}_i$ stands for the cosine of the angle between the handle and the rib $\mathbf{v}_i$, namely $\mathbf{c}^T \mathbf{v}_i = \cos(\alpha)$. Let $\beta = \pi/2$ be the angle between $\mathbf{v}_1$ and $\mathbf{v}_3$. The first spherical cosine theorem states that

$$\cos(\beta) = \cos^2(\alpha) + \sin^2(\alpha) \cos(\angle A).$$

Because angles α between the ribs and the handle are the same, the spherical angle $\angle A$ is the angle between the projection of the vectors $\mathbf{v}_1$ and $\mathbf{v}_3$ on the plane normal to the handle $\mathbf{c}$, i.e., $\angle A = 4\pi/5$. Finally, we can write

$$0 = \cos^2(\alpha) + \sin^2(\alpha) \cos(4\pi/5),$$

which gives $\cos^2(\alpha) = (\mathbf{c}^T \mathbf{v}_i)^2 = 1/\sqrt{5}$. Hence,

$$C_0(G_5) \le \log \vartheta(G_5) \le \log \left(\frac{1}{\cos^2(\alpha)} \right) = \log \sqrt{5} = \frac{1}{2} \log 5.$$

The opposite inequality is known and the Shannon lower bound is tight.

The definition of $\vartheta(G)$ is not unique. In his work, Lovász [7] pointed out four equivalent definitions for $\vartheta(G)$. For example, he showed that $\vartheta(G)$ is the minimum of the largest eigenvalue of any symmetric matrix $(a_{ij})_{i,j=1}^{|\mathcal{X}|}$ such that $a_{ij} = 1$ if $i = j$ or if x_i and x_j are non-adjacent.

Although the Lovász ϑ function behaves very beautifully, the value of $\log \vartheta(G)$ is generally different from the capacity. A new bound on the zero-error capacity was derived by Haemers [5], and it is sometimes better but quite often much worse than $\vartheta(G)$. A quadratic matrix of order $|\mathcal{X}|$ is said to *fit* the graph G if its diagonal entries are all nonzero and the element $a_{i,j}$ is zero if and only if the symbols x_i and x_j are adjacent in the channel. Haemers proved that the logarithm of the ranking of those matrices upper bounds the zero-error capacity of G. This result was illustrated with some examples for which his bound is better than $\vartheta(G)$. However, this is not true for the pentagon graph G_5.

The Lovász functional was generalized to the quantum scenario as an upper bound for the number of entanglement-assisted zero-error messages sent through a noisy quantum channel.

Next, we investigate the zero-error capacity of sum and product of discrete memoryless channels.

4.4 The Sum and Product of Channels

Consider two discrete memoryless channels $(\mathcal{X}_1, \mathcal{Y}_1, P_1)$ and $(\mathcal{X}_2, \mathcal{Y}_2, P_2)$ with zero-error capacities C_{0_1} and C_{0_2}, respectively. We are interested in transmitting information using the two channels and we ask for the zero-error capacity of the joint system [11]. Basically, there are two natural ways of assembling two channels to form a single channel, which we call the *sum* and the *product* of two channels.

The sum of two channels is a new channel $(\mathcal{X}_1 \sqcup \mathcal{X}_2, \mathcal{Y}_1 \sqcup \mathcal{Y}_2, P_1 \oplus P_2)$, where the stochastic matrix of the sum channel is the direct sum of the two stochastic matrices, and the input (output) set is the disjoint union of $\mathcal{X}_1$ ($\mathcal{Y}_1$) and $\mathcal{X}_2$ ($\mathcal{Y}_2$), respectively. Intuitively, the sum channel behaves as $(\mathcal{X}_1, \mathcal{Y}_1, P_1)$ if an input symbol $x_{1_i} \in \mathcal{X}_1$ is used. Otherwise, it behaves as $(\mathcal{X}_2, \mathcal{Y}_2, P_2)$. This corresponds physically to a situation where either of two channels may be used but not both. Analogously, the product channel is a new DMC $(\mathcal{X}_1 \times \mathcal{X}_2, \mathcal{Y}_1 \times \mathcal{Y}_2, P_1 \otimes P_2)$, where the stochastic matrix is the direct product of the two matrices, and the input (output) set is the Cartesian product of $\mathcal{X}_1$ ($\mathcal{Y}_1$) and $\mathcal{X}_2$ ($\mathcal{Y}_2$), respectively. In this case, we can think of the product DMC as of a nonstationary memoryless channel over which transmission is governed in strict alternation by the stochastic matrices P_1 and P_2:

$$p(y_{1_i}, y_{2_i} | x_{1_i}, x_{2_i}) = p(y_{1_i} | x_{1_i}) p(y_{2_i} | x_{2_i}).$$

Consider two DMCs, $(\mathcal{X}_1, \mathcal{Y}_1), P_1, (\mathcal{X}_2, \mathcal{Y}_2, P_2)$, and let C_1, C_2 be their corresponding ordinary capacities. It is well known [10] that the ordinary capacity of the sum channel is $C_+ = \log(\exp C_1 + \exp C_2)$. For the product channel, the ordinary capacity is proved to be $C_\times = C_1 + C_2$.

The error-free communication capacity of the sum and product channels was studied by Shannon [11]. If C_{0+} and $C_{0\times}$ denote the zero-error capacity of the sum and product channels, respectively, then Shannon demonstrated that

$$C_{0+} \geq \log(\exp C_{0_1} + \exp C_{0_2}) \tag{4.12}$$

and

$$C_{0\times} \geq C_{0_1} + C_{0_2}, \tag{4.13}$$

with equality if and only if the adjacent graph G of either of the two channels can be colored using $\alpha(G)$ colors. In an analogy with the ordinary capacity, Shannon conjectured that, in fact, equalities always hold for zero-error capacities. The product channel conjecture was implicitly disproved in a example of Haemers [5]. More recently, Alon [1] proved the existence of channels for which $C_{0\times} > C_{0_1} + C_{0_2}$. Such results suggest that the zero-error capacity behaves quite different from the ordinary capacity.

Suppose that $C_{0_1} = C_{0_2} = 0$. It is straightforward to see that $C_{0_+} = C_{0_\times} = 0$. In the quantum scenario, however, quantum effects can be used to *activate* the capability of two quantum channels perform, together, classical zero-error communication, even if both quantum channels have zero-error capacities equal to zero!

4.5   Further Reading

We have presented in this chapter fundamental concepts in zero-error information theory. We have started by defining the zero-error capacity of a DMC and introducing a method to calculate the capacity of simple channels. Then, the problem of finding the zero-error capacity was reformulated in terms of graph theory. It was shown how several results in zero-error theory can be restated in a graph language. The most famous upper bound on the zero-error capacity, the Lovász ϑ function, was presented and used to calculate the zero-error capacity of the pentagon graph, a problem that remained open during more than 20 years. Finally, we presented the zero-error capacity of sum and products of discrete memoryless channels.

The zero-error capacity of DMC was first studied by Shannon in 1956 [10]. Lovász introduced the theta functional in 1979 [7]. A wide survey about developments in the zero-error information theory until 1998 can be found in Körner and Orlitsky [6], as part of an IEEE Transactions on Information Theory commemorative issue due to the 50th anniversary of Shannon Theory.

References

1. Alon N (1998) The Shannon capacity of a union. Combinatorica 18(3):301–310. Accessed 28 Nov 2016
2. Bollobás B (1998) Modern graph theory. Springer, New York
3. Grötschel M, Lovász L (1988) Geometric algorithms and combinatorial optimization. Springer, Berlin
4. Haemers W (1978) An upper bound for the Shannon capacity of a graph. In: Algebraic methods in graph theory. Colloquium Mathematical Society, Szeged, Hungary, pp 267–272
5. Haemers W (1979) On some problems of Lovász concerning the Shannon capacity of a graph. IEEE Trans Inf Theory 25:231–232
6. Körner J, Orlitsky A (1998) Zero-error information theory. IEEE Trans Inf Theory 44(6):2207–2229
7. Lovász L (1979) On the Shannon capacity of a graph. IEEE Trans Inf Theory 25(1):1–7
8. Rosenfeld M (1967) On a problem of Shannon. Proc Am Math Soc 18:318–319
9. Scheinerman ER, Ullman DH (1997) Fractional graph theory. Wiley-Interscience, New York
10. Shannon CE (1948) A mathematical theory of communication. The Bell Syst Tech J 27:379–423, 623–656
11. Shannon CE (1956) The zero error capacity of a noisy channel. IRE Trans Inf Theory 2(3):8–19
12. van Lint JH, Wilson RM (2001) A course in combinatorics, 2nd edn. Cambridge University Press, Cambridge

Chapter 5
Zero-Error Capacity of Quantum Channels

Quantum zero-error information theory deals with techniques, protocols and analyzes methods to allow communication of classical or quantum information through quantum noisy channels, with the main requirement that no errors can be tolerated. This new research field aims to generalize and extend the classical zero-error theory proposed by Shannon and outlined in Chap. 4.

Because quantum mechanics has many features not present in the classical world, e.g. entanglement, one may expect that developments in this area should not only generalize definitions and results from classical theory but they must go beyond. This is exactly what has been happening! As we already know, a classical channel has an asymptotically positive zero-error capacity if and only if the one-shot capacity is positive, i.e., $N(1) > 1$. One of the most impressive results in quantum zero-error information theory is that there exist quantum channels such that no information can be perfectly transmitted with a single use, whereas the communication is possible with two channel uses. This phenomenon is known as superactivation of the quantum zero-error capacity.

In this chapter we present the classical zero-error capacity of quantum channels, which is a generalization of the zero-error capacity of discrete memoryless channels. Section 5.1 deals with main concepts and definitions. In Sect. 5.2, an equivalent definition for the quantum zero-error capacity is given in terms of graphs. Some properties of quantum states reaching the channel capacity are investigated in Sect. 5.3. Section 5.4 presents an upper bound for the quantum zero-error capacity in terms of the Holevo-Schumacher-Westmoreland capacity and, finally, Sect. 5.5 discusses the superactivation of the zero-error capacity of quantum channels.

© Springer International Publishing Switzerland 2016
E.B. Guedes et al., *Quantum Zero-Error Information Theory*,
DOI 10.1007/978-3-319-42794-2_5

5.1 Classical Zero-Error Capacity of Quantum Channels

In quantum information theory, the Holevo-Schumacher-Westmoreland (HSW) capacity can be understood as a quantum version of the ordinary capacity of classical channels originally defined by Shannon. Fundamentally, the parallel is due to the communication protocol used by the HSW capacity: quantum codewords are composed by tensor product of quantum states and collective measurements can be made at the channel output.

The first zero-error capacity of quantum channels was defined taking into account the HSW communication protocol, with the restriction that the probability of decoding errors should be zero. In this sense, the classical zero-error capacity (CZEC) of a quantum channel can be viewed as the "error-free" version of the HSW capacity, as well as the quantum generalization of the zero-error capacity of classical channels.

Given an arbitrary quantum channel, we ask for the maximum amount of classical information per channel use that can be transmitted with zero probability of error. For this purpose, we consider a d-dimensional quantum channel $\mathcal{E} \equiv \{E_a\}$ as a completely positive tracing preserving map. Let S be a subset of quantum states belonging to a d-dimensional Hilbert space $\mathcal{H}$. States $\rho_i \in S$ will be called *input states*.

Initially, a sender *Alice* chooses a message from a set $\{1, \ldots, m\}$ containing m classical messages. The encoder maps this message into an n-tensor product of quantum states of S. The resulting state is called *quantum codeword*, which is sent through the noisy quantum channel $\mathcal{E}$. The receiver *Bob* performs a collective measurement with a POVM on the received state. The measurement output becomes argument of the decoding function. The decoder has to decide which message was sent by Alice considering that errors are not allowed. The sketch of the communication protocol is shown in Fig. 5.1.

The error-free communication protocol can be summarized as follows.

- The source alphabet is the set $S = \{\rho_1, \ldots, \rho_\ell\}$, where $S \subseteq \mathcal{H}$;
- In order to be transmitted through the quantum channel, classical messages are mapped into tensor products of quantum states in S;

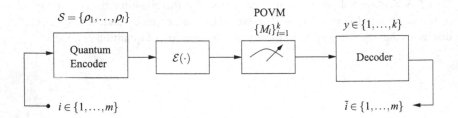

Fig. 5.1 Quantum zero-error communication system

- At the channel output, the decoder performs collective measurements in order to estimate the message that was sent.

Taking this into account, we can now define error-free quantum codes.

Definition 5.1 ((m, n) Error-Free Quantum Code). An (m, n) error-free quantum code for a quantum channel $\mathcal{E}$ is composed of:

1. a set of indexes $\{1, \ldots, m\}$, each one associated with a classical message;
2. an encoding function

$$f_n : \{1, \ldots, m\} \to \mathcal{S}^{\otimes n} \tag{5.1}$$

leading to codewords $f_n(1) = \overline{\rho}_1, \ldots, f_n(m) = \overline{\rho}_m, \overline{\rho}_i \in \mathcal{S}^{\otimes n}$;
3. a decoding function

$$g : \{1, \ldots, k\} \to \{1, \ldots, m\} \tag{5.2}$$

that deterministically associates each of the possible measurement results performed by a POVM $\mathcal{M} = \{M_i\}_{i=1}^{k}$ with a message. The decoding function has the following property:

$$\Pr[g(\mathcal{E}(f_n(i))) \neq i] = 0 \qquad \forall i \in \{1, \ldots, m\}. \tag{5.3}$$

Clearly, the rate of this code is $R_n = \frac{1}{n} \log m$ bits per channel use.

With these codes we can now define the *classical zero-error capacity* (CZEC) of a quantum channel .

Definition 5.2 (Quantum Zero-Error Capacity [11]). Let $\mathcal{E}(\cdot)$ be a TPCP map that represents a quantum noisy channel. The quantum zero-error capacity of $\mathcal{E}$, denoted by $C^{(0)}(\mathcal{E})$, is the highest superior limit of achievable rates with zero-error decoding probability, i.e,

$$C^{(0)}(\mathcal{E}) = \sup_{\mathcal{S}} \sup_{n} \frac{1}{n} \log \alpha_n(\mathcal{E}), \tag{5.4}$$

where $\alpha_n(\mathcal{E}) = m$ is the maximum number of classical messages that can be transmitted without errors when an (m, n) error-free quantum code is used with input alphabet $\mathcal{S}$.

For a given (m, n) error-free quantum code attaining the quantum zero-error capacity of $\mathcal{E}$, we define an optimum pair $(\mathcal{S}, \mathcal{M})$.

Definition 5.3 (Optimum Pair $(\mathcal{S}, \mathcal{M})$ [15]). An optimum pair $(\mathcal{S}, \mathcal{M})$ is composed by a set of input states $\mathcal{S}$ and a POVM $\mathcal{M}$ for which the quantum zero-error capacity is reached.

We are interested in scenarios where the classical zero-error capacity of a quantum channel is non-zero. This is only possible when at least two input states are non-adjacent. Definition 5.4 synthesizes this idea.

Definition 5.4 (Adjacency of Quantum States). Let $\mathcal{E}$ be a quantum channel and $\rho_i, \rho_j \in S$ two quantum states, $i \neq j$, from the input alphabet of the channel. We say that ρ_i and ρ_j are *non-adjacent* (orthogonal or distinguishable) in the output of $\mathcal{E}$ if the Hilbert subspaces spanned by the supports of ρ_i and ρ_j are orthogonal. We denote it by $\rho_i \perp_{\mathcal{E}} \rho_j$.

Adjacency of quantum states can be generalized to tensor product, as illustrated in Fig. 5.2. Let $\hat{\rho}_i$ and $\hat{\rho}_j$ be two input tensor products $\hat{\rho}_i = \rho_{i,1} \otimes \ldots \otimes \rho_{i,n}$ and $\hat{\rho}_j = \rho_{j,1} \otimes \ldots \otimes \rho_{j,n}$. If there is at least one $\rho_{i,k} \perp_{\mathcal{E}} \rho_{j,k}$, then $\hat{\rho}_i \perp_{\mathcal{E}} \hat{\rho}_j$.

Taking adjacency into account, a quantum channel $\mathcal{E}$ has positive zero-error capacity if and only if the set S contains at least two non-adjacent states.

Example 5.1 (Quantum Channel with a Vanishing Zero-Error Capacity). Suppose that a depolarizing quantum channel can transmit an input state ρ intact with probability $1 - p$ or exchange it by a complete mixed state with probability p, as discussed in Example 3.15. Let d be the dimension of the input Hilbert state $\mathcal{H}$ and let $\mathbb{1}_d$ be the identity matrix of dimension d. This channel is shown in Fig. 5.3.

Recall that the formal representation of this channel is given by

$$\mathcal{E}(\rho) = (1 - p)\rho + p\frac{1}{d}\mathbb{1}_d, \tag{5.5}$$

where $0 < p < 1$. To check if this channel has positive zero-error capacity, we verify if there exist at least two different states, ρ_i and ρ_j, which are distinguishable at the channel output, i.e.,

$$\text{Tr}[\mathcal{E}(\rho_i)\mathcal{E}(\rho_j)] = \text{Tr}\left[\left((1 - p)\rho_i + p\frac{1}{d}\mathbb{1}_d\right)\left((1 - p)\rho_j + p\frac{1}{d}\mathbb{1}_d\right)\right]$$

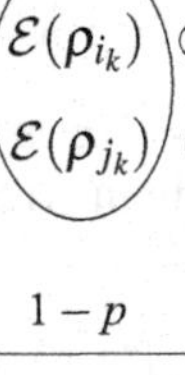

Fig. 5.2 Two quantum states $\hat{\rho}_i$ e $\hat{\rho}_j$ are distinguishable if there is at least one $\rho_{i,k} \perp_{\mathcal{E}} \rho_{j,k}, 1 \leq k \leq n$

$$\mathcal{E}(\hat{\rho}_i) = \mathcal{E}(\rho_{i_1}) \otimes \cdots \otimes \left(\mathcal{E}(\rho_{i_k})\right) \otimes \cdots \otimes \mathcal{E}(\rho_{i_n})$$
$$\mathcal{E}(\hat{\rho}_j) = \mathcal{E}(\rho_{j_1}) \otimes \cdots \otimes \left(\mathcal{E}(\rho_{j_k})\right) \otimes \cdots \otimes \mathcal{E}(\rho_{j_n})$$

Fig. 5.3 Quantum depolarizing channel

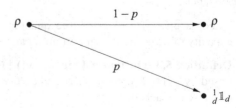

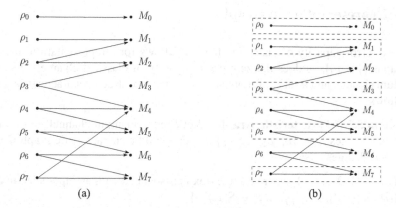

Fig. 5.4 Diagrams representing the transitions performed by the channel ε over the input. (a) A discrete memoryless channel. (b) A subset of non-adjacent states at the channel output

$$= \mathrm{Tr}\left[(1-p)^2 \, \mathrm{Tr}[\rho_i \rho_j] + \frac{p(1-p)}{d} \, \mathrm{Tr}[\rho_i + \rho_j] + \frac{p^2}{d^2} \mathbb{1}_d\right]$$
$$> 0,$$

for any $0 < p < 1$. This way, the zero-error capacity of the depolarizing quantum channel $\mathcal{E}$ is zero, i e , $C^{(0)}(\mathcal{E}) = 0$.

Example 5.2 (Quantum Channel with Positive Zero-Error Capacity). Suppose a quantum channel $\mathcal{E}$ in an 8-dimensional Hilbert space. Consider a set of classical messages $\{0, 1, \ldots, 7\}$ associated with a set of pure quantum input states $\mathcal{S} = \{\rho_0 = |0\rangle \langle 0|, \rho_1 = |1\rangle \langle 1|, \ldots, \rho_7 = |7\rangle \langle 7|\}$, in which $0 \mapsto \rho_0, 1 \mapsto \rho_1, \ldots, 7 \mapsto \rho_7$. Let $\mathcal{M}$ be a POVM specified by $\mathcal{M} = \{M_i = |i\rangle \langle i|\}_{i=0}^{7}$. Notice that $\sum_{i=0}^{7} M_i = \mathbb{1}$. We consider that the quantum channel acts on the input as shown in Fig. 5.4a.

This channel has positive zero-error capacity because it is possible to identify a subset of non-adjacent states at the channel output, as shown in Fig. 5.4b. This subset is composed by $\{\rho_0, \rho_1, \rho_3, \rho_5, \rho_7\}$ and it is maximal. We have that

$$C^{(0)}(\mathcal{E}) \geq \frac{1}{1} \log_2 5$$

$$\geq 2.321 \text{ bits per symbol per channel use.}$$

It is important to emphasize that it is not possible to state that the zero-error capacity of this quantum channel is equal to 2.321 bits per channel use, since the supremum (5.4) should be taken over all sets $\mathcal{S}$ and over all zero-error quantum codes of length n.

5.2   Representation in Graphs

The zero-error capacity of quantum channels allows for a representation in terms of graphs [11, 14], as does the zero-error capacity of classical channels. Given a quantum channel and a set S of input states, we can construct a characteristic graph as follows.

Definition 5.5 (Characteristic Graph). Let $\mathcal{E}$ be a quantum channel. For a given set of input states $S = \{\rho_1, \rho_2, \ldots, \rho_\ell\}$, we can build a characteristic graph $\mathcal{G}$ with vertices and edges given by

1. $V(\mathcal{G}) = \{1, 2, \ldots, \ell\}$, where each vertex is associated with an input state in S;
2. $E(\mathcal{G}) = \{(i, j) | \rho_i \perp_{\mathcal{E}} \rho_j; \rho_i, \rho_j \in S; i \neq j\}$.

This notion of characteristic graph can be extended to the n-tensor product of states in S, $S^{\otimes n}$, giving rise to the graph $\mathcal{G}^n$, where $V(\mathcal{G}^n) = V(\mathcal{G})^n$ and two vertices in $V(\mathcal{G}^n)$ are connected if and only if the corresponding n-tensor input states are non-adjacent at the channel output, i.e.,

1. $V(\mathcal{G}^n) = \{1, 2, \ldots, \ell\}^n$,
2. $E(\mathcal{G}^n) = \{(i_1 \ldots i_n, j_1 \ldots j_n) | \rho_{i_k} \perp_{\mathcal{E}} \rho_{j_k}\}$ for at least one k, $1 \leq k \leq n$; $\rho_{i_k}, \rho_{j_k} \in S$.

With this representation we can verify that vertices connected by an edge in the graph $\mathcal{G}^n$ correspond to mutually non-adjacent product states at the channel output. Therefore, the maximum amount of messages that can be transmitted by an (m, n) error-free quantum code with input alphabet S is given by the clique number of $\mathcal{G}^n$, denoted by $\omega(\mathcal{G}^n)$, i.e., $\omega(\mathcal{G}^n) = \alpha_n(\mathcal{E})$. Moreover, we can give an alternative, equivalent definition of the zero-error capacity of quantum channels in terms of graph theory.

Definition 5.6 (Classical Zero-Error Capacity of a Quantum Channel). The zero-error capacity of a quantum channel $\mathcal{E}$ is given by

$$C^{(0)}(\mathcal{E}) = \sup_{S} \sup_{n} \frac{1}{n} \log \omega(\mathcal{G}^n), \tag{5.6}$$

where the supremum is taken over all input sets S and over all codes of length n.

Example 5.3 (Zero-Error Capacity from a Graph). Let's return to the channel $\mathcal{E}$ of Example 5.2 and illustrated in Fig. 5.4a. The characteristic graph associated with the subset S is shown in Fig. 5.5. The clique number of this characteristic graph is 5, corresponding to the pairs $\{(0, 1), (1, 3), (3, 5), (5, 7), (7, 0)\}$. The clique of the characteristic graph of $\mathcal{E}$ is shown in Fig. 5.5 with dotted edges.

Fig. 5.5 Characteristic
Graph of $\mathcal{E}$ from Example 5.2

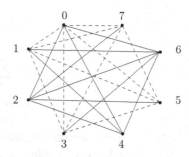

5.3   Quantum States Attaining the Zero-Error Capacity

We turn our attention to the subset $\mathcal{S} = \{\rho_1, \dots, \rho_l\}$ that reaches the supremum (5.6). Note that states $\rho_i \in \mathcal{S}$ can be either *pure* or *mixed* states. It turns out that the zero-error capacity of a quantum channel can always be reached by a set of quantum pure states.

Proposition 5.1. *The zero-error capacity of quantum channels can be achieved by a set $\mathcal{S}$ composed only of pure quantum states, i.e., $\mathcal{S} = \{\rho_i = |v_i\rangle\langle v_i|\}$.*

To prove the proposition, consider a quantum channel with Kraus operators $\mathcal{E} \equiv \{E_a\}$, and assume that the capacity is reached by a subset $\mathcal{S}$ that contains mixed states. Then, we show that it is always possible to write a new subset $\mathcal{S}'$ composed only of pure states that also achieves the channel capacity.

Initially, note that two quantum states ρ and σ have orthogonal supports if and only if $\mathrm{Tr}\,(\rho\sigma) = 0$. We can write

$$
\mathrm{Tr}\left(\mathcal{E}(\rho_i)\mathcal{E}(\rho_j)\right) = \mathrm{Tr}\left(\sum_a \sum_r \lambda_{i_r} E_a |v_{i_r}\rangle\langle v_{i_r}| E_a^\dagger \sum_b \sum_s \lambda_{j_s} E_b |v_{j_s}\rangle\langle v_{j_s}| E_b^\dagger\right)
$$

$$
= \mathrm{Tr}\left(\sum_a \sum_r \sum_b \sum_s \lambda_{i_r}\lambda_{j_s} E_a |v_{i_r}\rangle\langle v_{i_r}| E_a^\dagger E_b |v_{j_s}\rangle\langle v_{j_s}| E_b^\dagger\right)
$$

$$
= \sum_{a,r,b,s} \lambda_{i_r}\lambda_{j_s} ||\langle v_{i_r}| E_a^\dagger E_b |v_{j_s}\rangle||^2, \tag{5.7}
$$

where $\rho_i = \sum_s \lambda_{i_s} |v_{i_s}\rangle\langle v_{i_s}|$. Suppose that $\rho_i \perp_{\mathcal{E}} \rho_j$, then $\langle v_{i_r}| E_a^\dagger E_b |v_{j_s}\rangle = 0$ for all indexes r and s. Now, without loss of generality, define a new set $\mathcal{S}' = \{|v_{1_1}\rangle, \dots, |v_{l_1}\rangle\}$, where $|v_{i_1}\rangle \in \mathrm{supp}\,\rho_i$ is a pure state in the support of ρ_i. It is clear that if ρ_i and ρ_j are non-adjacent, then

$$
\mathrm{Tr}\left(\mathcal{E}(|v_{i_1}\rangle)\mathcal{E}(|v_{j_1}\rangle)\right) = \mathrm{Tr}\left(\sum_a E_a |v_{i_1}\rangle\langle v_{i_1}| E_a^\dagger \sum_b E_b |v_{j_1}\rangle\langle v_{j_1}| E_b^\dagger\right)
$$

$$= \mathrm{Tr}\left(\sum_a \sum_b E_a |v_{i_1}\rangle\langle v_{i_1}|E_a^\dagger E_b|v_{j_1}\rangle\langle v_{j_1}|E_b^\dagger\right)$$

$$= \sum_{a,b} ||\langle v_{i_1}|E_a^\dagger E_b|v_{j_1}\rangle||^2$$

$$= 0. \tag{5.8}$$

Note that all non-adjacency relationships in S are at least preserved when using the subset S'. In terms of graphs, the characteristic graph $\mathcal{G}'$ due to S' can be obtained from $\mathcal{G}$, the characteristic graph due to S, by probably adding a number of edges but never deleting edges. Because adding new vertices never decreases the clique number of a graph, we can conclude that the subset S' also attains the quantum zero-error capacity.

Now the relationship between orthogonality at the channel input and non-adjacency at the channel output is investigated. It is straightforward to see that two non-adjacent quantum states are necessarily orthogonal at the channel input. At a first glance, we can think that maximizing the number of pairwise non-adjacent quantum states requires pairwise orthogonal states at the channel input. Surprisingly, it turns out that there exist quantum channels such that we can do better by choosing a subset S where not all states are pairwise orthogonal.

To illustrate this feature, we present a mathematically motivated example of a quantum channel for which the capacity is attained by a set of non-orthogonal input states. In addition, the channel gives rise to the pentagon graph for the subset reaching the capacity.

Example 5.4. Let e be a quantum channel with operation elements $\{E_1, E_2, E_3\}$ given by

$$E_1 = \begin{bmatrix} 0.5 & 0 & 0 & 0 & \frac{\sqrt{49902}}{620} \\ 0.5 & -0.5 & 0 & 0 & 0 \\ 0 & 0.5 & -0.5 & 0 & 0 \\ 0 & 0 & 0.5 & -\frac{\sqrt{457}}{50} & \frac{\sqrt{457}}{50} \\ 0 & 0 & 0 & -0.62 & -\frac{289}{1550} \end{bmatrix}, \quad E2 = \begin{bmatrix} 0.5 & 0 & 0 & 0 & -\frac{\sqrt{49902}}{620} \\ 0.5 & 0.5 & 0 & 0 & 0 \\ 0 & 0.5 & 0.5 & 0 & 0 \\ 0 & 0 & 0.5 & \frac{\sqrt{457}}{50} & -\frac{\sqrt{457}}{50} \\ 0 & 0 & 0 & 0.5 & 0.5 \end{bmatrix},$$

$$E3 = 0.3|4\rangle\langle 4|,$$

where $\beta = \{|0\rangle, \ldots, |4\rangle\}$ is the computational basis for the Hilbert space of dimension five. Consider the following set S of input states for $\mathcal{E}$:

$$S = \left\{|v_1\rangle = |0\rangle, |v_2\rangle = |1\rangle, |v_3\rangle = |2\rangle, |v_4\rangle = |3\rangle, |v_5\rangle = \frac{|3\rangle + |4\rangle}{\sqrt{2}}\right\}. \tag{5.9}$$

In order to construct the characteristic graph $\mathcal{G}$, we need to explicit all adjacency relations between states in S. If the channel outputs $\mathcal{E}(|v_i\rangle)$ are calculated for every

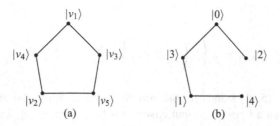

Fig. 5.6 (**a**) Characteristic graph $\mathcal{G}$ for the subset S containing non-adjacent input states. (**b**) Characteristic graph for a subset S' of mutually orthogonal input states. In this case, the transmission rate is less than $C^{(0)}$ (pentagon) for any zero-error quantum code with alphabet S'

$|v_i\rangle \in S$, one can verify that

$$|v_1\rangle \perp_{\mathcal{E}} |v_3\rangle, \qquad |v_1\rangle \perp_{\mathcal{E}} |v_4\rangle, \qquad |v_2\rangle \perp_{\mathcal{E}} |v_4\rangle,$$

$$|v_2\rangle \perp_{\mathcal{E}} |v_5\rangle, \qquad \text{and} \qquad |v_3\rangle \perp_{\mathcal{E}} |v_5\rangle.$$

These relations give rise to the pentagon as characteristic graph, as it is illustrated in Fig. 5.6a.

It is important to note that if we replace the state $|v_5\rangle = \frac{|3\rangle + |4\rangle}{\sqrt{2}}$ with the state $|4\rangle$, then the subset S becomes the (orthonormal) basis β. For the subset β, the characteristic graph is shown in Fig. 5.6b. The zero-error capacity of the former graph is $C_0 = \frac{1}{2}\log 5$ bits/use, whereas the latter has zero-error capacity $C_0 = 1$ bit/use. Because there is no other subset of input states giving rise to a graph with $C_0 \geq 1$, the zero-error capacity of $\mathcal{E}$ is

$$C^{(0)}(\mathcal{E}) = \frac{1}{2} \log 5 \text{ bits/use.}$$

5.4   Relation with Holevo-Schumacher-Westmoreland Capacity

Quantum channels have different kinds of capacity, depending mainly on if the information to be sent is either classical or quantum and on the communication protocol [17]. When classical messages are mapped into tensor products at the channel input and collective measurements are performed at the channel output, the capacity of the quantum channel to convey classical information with a negligible error probability after many channel uses, denoted by $C_{1,\infty}$, is given by the *Holevo-Schumacher-Westmoreland* (HSW) *theorem* [10, 20].

As we already mentioned, the HSW capacity can be understood as a generalization of the ordinary capacity of classical channels. According to the HSW theorem, this capacity is given by

$$C_{1,\infty}(\mathcal{E}) \equiv \max_{p_i, \rho_i} \chi_{p_i, \rho_i}, \tag{5.10}$$

where

$$\chi_{p_i,\rho_i} = S\left(\mathcal{E}\left(\sum_i p_i\rho_i\right)\right) - \sum_i p_i S\left(\mathcal{E}(\rho_i)\right). \tag{5.11}$$

The term $S(\cdot)$ in (5.11) stands for the von Neumann entropy; the maximum (5.10) takes into account all possible input ensembles $\{\rho_i, p_i\}$ for the quantum channel $\mathcal{E}$; χ_{p_i,ρ_i} is known as the *Holevo quantity*.

Theorem 5.1 (Bound on the Quantum Zero-Error Capacity). *The zero-error capacity of a quantum channel $\mathcal{E}$ is upper bounded by the HSW capacity, i.e.,*

$$C^{(0)}(\mathcal{E}) \leq C_{1,\infty}(\mathcal{E}) \equiv \max_{p_i,\rho_i} \chi_{p_i,\rho_i}. \tag{5.12}$$

To prove the theorem, we assume that Alice sends to Bob a message chosen uniformly from the set $\{1, \ldots, 2^{nR}\}$. If we define a random variable X representing indexes of classical messages, then

$$H(X) = nR, \tag{5.13}$$

where H stands for the classical Shannon entropy [1]. Let Y be a random variable representing POVM outputs. Using the definition of mutual information, we get

$$nR = H(X) = H(X|Y) + I(X,Y). \tag{5.14}$$

Because the quantum code is error-free, $H(X|Y) = 0$. Suppose that Alice encodes the message i as $\overline{\rho}_i = \rho_{i_1} \otimes \cdots \otimes \rho_{i_n}$. Applying the Holevo bound we get

$$nR = I(X,Y)$$

$$\leq S\left(\sum_{i=1}^{2^{nR}} \frac{1}{2^{nR}} \mathcal{E}(\overline{\rho}_i)\right) - \sum_{i=1}^{2^{nR}} \frac{1}{2^{nR}} S(\mathcal{E}(\overline{\rho}_i)). \tag{5.15}$$

Remember that $\mathcal{E}(\overline{\rho}_i) = \mathcal{E}(\rho_{i_1}) \otimes \cdots \otimes \mathcal{E}(\rho_{i_n})$. Hence, we can apply the subadditivity of the von Neumann entropy, $S(A,B) \leq S(A) + S(B)$:

$$nR \leq \sum_{j=1}^{n} S\left(\sum_{i=1}^{2^{nR}} \frac{1}{2^{nR}} \mathcal{E}(\rho_{i_j})\right) - \sum_{i=1}^{2^{nR}} \frac{1}{2^{nR}} \sum_{j=1}^{n} S(\mathcal{E}(\rho_{i_j}))$$

$$= \sum_{j=1}^{n} \left[S\left(\sum_{i=1}^{2^{nR}} \frac{1}{2^{nR}} \mathcal{E}(\rho_{i_j})\right) - \sum_{i=1}^{2^{nR}} \frac{1}{2^{nR}} S(\mathcal{E}(\rho_{i_j})) \right]. \tag{5.16}$$

Because the capacity (5.10) is calculated by taking the ensemble that gives the maximum, we can conclude that each term on the right side of (5.16) is less than or equal to $C_{1,\infty}(\mathcal{E})$. Then,

$$nR \leq nC_{1,\infty}(\mathcal{E}) \tag{5.17}$$

and the inequality follows for all zero-error quantum block codes of length n and rate R. This is an intuitive result, since one would expect to increase the information transmission rate whenever a small probability of error is allowed.

Example 5.5. Consider the quantum channel of the Example 5.4 and the set $\mathcal{S}$ of non-orthogonal states giving rise to the pentagon as characteristic graph. Obviously, we do not know if $\mathcal{S}$ attains the supremum (5.10). However, $\mathcal{S}$ does attain the zero-error capacity of $\mathcal{E}$, which is $C_0(G_5) = \frac{1}{2}\log 5$ bits per use. In this case, a simple calculation shows that the χ quantity for the family $\{\mathcal{S}, p_i = 1/5\}$ is greater than $C_0(G_5)$, i.e.,

$$\chi_{\{\mathcal{S},1/5\}} = \frac{1}{5}\left[S\left(\mathcal{E}\left(\sum_{i=1}^{5}|v_i\rangle\langle v_i|\right)\right) - \sum_{i=1}^{5} S(\mathcal{E}(|v_i\rangle\langle v_i|))\right]$$

$$= 1.53$$

$$\geq C_0(G_5)$$

$$= 1.16. \tag{5.18}$$

5.5 Superactivation of Zero-Error Capacity

A well-known result in classical zero-error communication asserts that a discrete memoryless channel has positive zero-error capacity if and only if $N(1) > 1$. Therefore, if one use of the channel cannot transmit zero-error information, then many uses cannot either. Thanks to entanglement, the capacity of quantum channels to carry classical or quantum information behaves significantly different from the corresponding classical capacity. Concerning the zero-error capacity, there are quantum channels such that the use of entangled input states allows for a positive zero-error capacity even when the one-shot capacity is zero.[1] This phenomenon is known as *superactivation* of zero-error capacity of quantum channels.

The activation of zero-error capacity was first demonstrated by Duan and Shi [8] in a scenario of a multipartite communication system, where m senders want to send classical messages to n receivers using a noisy multipartite quantum channel.

---

[1] By one-shot capacity we mean the transmission rate for a single channel use.

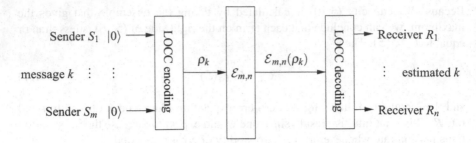

Fig. 5.7 An (m,n) multipartite quantum channel. When the senders want to transmit the message k, they start with the state $|0\rangle \otimes \cdots \otimes |0\rangle$. Then, LOCC is used to encode $|0\rangle^{\otimes m}$ into a quantum stated ρ_k, which is transmitted through the channel $\mathcal{E}_{m,n}$. At the channel output, receivers make use of LOCC to decode $\mathcal{E}_{m,n}(\rho_k)$ and get an estimation of the original message k

Afterwards, superactivation was found on several classes of one sender/one receiver quantum channels.

5.5.1 Activation of Zero-Error Capacity on Multipartite Quantum Channels

Consider a scenario where n senders, $S_1, \ldots, S_n$, want to communicate with m receivers, $R_1, \ldots, R_m$, using a multipartite quantum communication channel, $\mathcal{E}$. Figure 5.7 illustrates the general setup of this system. As a reasonable assumption, Duan and Shi considered that the senders can use LOCC (local operations and classical communication) in order to prepare and code the quantum state to be transmitted [8]. At the channel output, the receivers can also use LOCC to decode the output message.

Let $\mathcal{E}$ be an (m,n) multipartite quantum channel defined as the following positive trace-preserving map:

$$\mathcal{E}_{m,n} : \mathcal{B}(\mathcal{H}_S) \longrightarrow \mathcal{B}(\mathcal{H}_R),$$

where $\mathcal{H}_S = \mathcal{H}_{S_1} \otimes \cdots \otimes \mathcal{H}_{S_m}$ and $\mathcal{H}_R = \mathcal{H}_{R_1} \otimes \cdots \otimes \mathcal{H}_{R_n}$ are the state spaces of senders and receivers, respectively. In order to transmit a message, the senders start with a state $|0\rangle \otimes \cdots \otimes |0\rangle$ and prepare the input quantum codeword $\rho \in \mathcal{H}_S$ using LOCC. At the channel output, the receivers make use of LOCC to decode the output state $\mathcal{E}_{m,n}(\rho)$. The communication scheme is illustrated in Fig. 5.8.

As an example, define $\sigma_0 = |\beta_{00}\rangle\langle\beta_{00}|$ and $\sigma_1 = \frac{1}{3}(\mathbb{1} - \sigma_0)$, where $|\beta_{00}\rangle = \frac{|00\rangle+|11\rangle}{\sqrt{2}}$ stands for the Bell state. Consider the following one sender (Charlie) two receivers (Alice an Bob) quantum channel

$$\mathcal{E}_{1,2}(\rho) = \langle 0|\rho|0\rangle\sigma_0 + \langle 1|\rho|1\rangle\sigma_1, \tag{5.19}$$

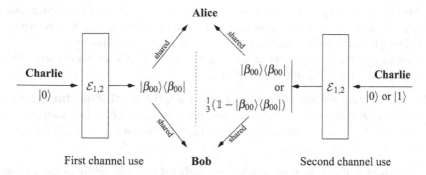

Fig. 5.8 Example of a $\mathcal{E}_{1,2}$ channel whose zero-error capacity can be activated. Although no perfect transmission can be archived with a single use, a sender can transmit one bit of information by using the channel twice. The zero-error capacity of this channel is at least 0.5 bits per channel use

which takes a qubit ρ into two qubits—one qubit for each receiver. It is straightforward to see that there exists just one pair of adjacent input states for this channel, corresponding to the qubits $|0\rangle$ and $|1\rangle$, since $\mathcal{E}_{1,2}(|0\rangle\langle0|) = \sigma_0$ and $\mathcal{E}_{1,2}(|1\rangle\langle1|) = \sigma_1$. Although σ_0 and σ_1 are orthogonal quantum states, Alice and Bob are not able to distinguish them after a single use of the channel. This arises from the fact that no quantum communication is allowed between the receivers. Because the one-shot zero error capacity of this channel is zero, one may think that no zero-error information can be transmitted, even after many channel uses. Shannon proved that this assertion is always true for classical channels. Thanks to entanglement, quantum channels behave drastically different from classical channels. Now suppose that the Charlie uses the channel $\mathcal{E}_{1,2}(\rho)$ twice to transmit $\rho^{\otimes 2} = |00\rangle$ or $\rho^{\otimes 2} = |01\rangle$. The corresponding received states are

$$\mathcal{E}_{1,2}^{\otimes 2}(|00\rangle\langle00|) = \sigma_0 \otimes \sigma_0, \tag{5.20}$$

$$\mathcal{E}_{1,2}^{\otimes 2}(|01\rangle\langle01|) = \sigma_0 \otimes \sigma_1. \tag{5.21}$$

No matter what are the messages transmitted by Charlie, $|00\rangle$ or $|01\rangle$, Alice and Bob always will share the Bell state $\sigma_0 = |\beta_{00}\rangle$. In order to complete the communication, Alice and Bob make use of the shared state $|\beta_{00}\rangle$ to teleport the second qubit, e.g., Alice teleports his part of the entangled state σ_0 or σ_1 to Bob. Finally, Bob performs a projective measurement in order to distinguish between σ_0 or σ_1. Because we are able to transmit two messages without confusion using the channel twice, the zero-error capacity is, at least,

$$C^{(0)}(\mathcal{E}_{1,2}) \geq \frac{1}{2}\log 2 = 0.5 \text{ bits per channel use.}$$

This is an amazing result with no counterpart in classical zero-error information theory. A single use of $\mathcal{E}_{1,2}$ cannot transmit error-free messages, whereas two uses can! This phenomenon is called *activation* of the zero-error capacity.

Considering that senders and receivers agree on the LOCC protocol above, Duan and Shi proved the following theorem:

Theorem 5.2 (Activation of the Zero-Error Capacity on Multipartite Quantum Channels). *For any $m > 1$ or $n > 1$, there exist (m, n) multipartite quantum channels for which one use of channel cannot transmit zero-error classical information, whereas two or more uses can.*

To prove the theorem, it is sufficient to explicit two multipartite quantum channels, $\mathcal{E}_{1,2}$ and $\mathcal{E}_{2,1}$, for which the zero-error capacity can be activated. That is because any (m, n) multipartite quantum channels can be extended to an $(m + m', n + n')$ channel, m' and n' positive integers, by ignoring the input from the additional m' senders and setting to $|0\rangle$ the output to all the additional n' receivers.

Consider a $(2, 1)$ multipartite quantum channel $\mathcal{E}_{2,1}$ from two senders, Alice and Bob, to one receiver, Charlie,

$$\mathcal{E}_{2,1} : \mathcal{B}(\mathcal{H}_S) \longrightarrow \mathcal{B}(\mathcal{H}_R), \tag{5.22}$$

where $\mathcal{H}_S = \mathcal{H}_{S_A} \otimes \mathcal{H}_{S_B}$ and $\mathcal{H}_R = \mathcal{H}_{R_C}$. The state spaces of Alice and Bob, $\mathcal{H}_{S_A}$ and $\mathcal{H}_{S_B}$, are four dimensional spaces. The output state space $\mathcal{H}_{R_C}$ is a qubit. The quantum channel $\mathcal{E}_{2,1}$ is defined as follows:

$$\mathcal{E}_{2,1}(\rho) = \mathrm{Tr}\,(P_0\rho)\,|0\rangle\langle 0| + \mathrm{Tr}\,(P_1\rho)\,|1\rangle\langle 1|, \tag{5.23}$$

where P_0 is a projector onto the state space $\mathcal{S}_0 \subset \mathcal{H}_S$ spanned by the (unnormalized) vectors

$$
\begin{aligned}
|\psi_1\rangle &= |00\rangle - |11\rangle, \\
|\psi_2\rangle &= |22\rangle - |33\rangle, \\
|\psi_3\rangle &= |20\rangle - |31\rangle, \\
|\psi_4\rangle &= |02\rangle + |13\rangle, \\
|\psi_5\rangle &= |30\rangle - |03\rangle, \\
|\psi_6\rangle &= |10\rangle - \sqrt{2}|21\rangle + |32\rangle, \\
|\psi_7\rangle &= |01\rangle + \sqrt{2}|12\rangle + |23\rangle, \\
|\psi_8\rangle &= |10\rangle - |32\rangle - |01\rangle + |23\rangle,
\end{aligned}
\tag{5.24}
$$

and P_1 is the projector onto $\mathcal{S}_1$, which is the orthogonal complement of $\mathcal{S}_0$, i.e., $\mathcal{S}_1 = \mathcal{S}_0^\perp$. The vectors $|\psi_i\rangle$ were carefully chosen in order to span a completely entangled state subspace. Consequently, $\mathcal{S}_0$ has no product state, which means that

S_0 does not have any state $|\phi\rangle$ such that $|\phi\rangle = |\phi_A\rangle \otimes |\phi_B\rangle$, $|\phi_A\rangle \in \mathcal{H}_{S_A}$ and $|\phi_B\rangle \in \mathcal{H}_{S_B}$. Because any quantum codeword ρ prepared by Alice and Bob using LOCC is necessarily a product state, it turns out that $\text{Tr}(P_0\rho) > 0$ as well as $\text{Tr}(P_1\rho) > 0$. As a consequence, the corresponding outputs of the channel (5.23) are always non-orthogonal mixed states. Therefore, there are no pairs of adjacent states at the channel input and the one-shot zero-error capacity is zero, i.e., no classical information can be transmitted with a single use of the $\mathcal{E}_{2,1}$ channel. However, the use of entanglement between two uses of channel enables the transmission of a classical bit with a zero probability of error.

Let $|\Phi\rangle$ be the following bipartite quantum state:

$$|\Phi\rangle = \frac{1}{2}(|00\rangle + |11\rangle + |22\rangle + |33\rangle). \tag{5.25}$$

Because $|\Phi\rangle \in \mathcal{H}_S \otimes \mathcal{H}_S$, we denote by $|\Phi\rangle_{AA'}$ the multipartite state $|\Phi\rangle$ prepared by Alice, where A and A' denote the first and the second component of $|\Phi\rangle$, respectively. The same holds for the state $|\Phi\rangle_{BB'}$ prepared by Bob. Define $U_i = |0\rangle\langle0| - |1\rangle\langle1| + |2\rangle\langle2| - |3\rangle\langle3|$ as an operator that acts on the component i of $|\Phi\rangle$, where $i \in \{A, A', B, B'\}$. In order to activate the zero-error capacity of the channel, the senders use the quantum channel twice in the following way:

1. Alice locally prepares the state $|\Phi\rangle$, denoted by $|\Phi\rangle_{AA'}$. Bob does the same, getting $|\Phi\rangle_{BB'}$;
2. Using LOCC, Alice and Bob agree on who will transmit the message (one bit) to Charlie. Without loss of generality, suppose that Alice sends the message (the bit "0") to Charlie;
3. Alice and Bob transmit the first components of their bipartite state, i.e., the components A and B of $|\Phi\rangle_{AA'}$ and $|\Phi\rangle_{BB'}$, respectively. This is the first use of the channel;
4. The second components of each state $|\Phi^A\rangle$ and $|\Phi\rangle_{BB'}$ are sent;
5. After the second round, Charlie performs a joint projective measurement in order to estimate the message sent by Alice and Bob. As will be explained below, the received state is given by

$$\mathcal{E}_{2,1}^{\otimes2}(|\Phi\rangle_{AA'} \otimes |\Phi\rangle_{BB'}) = \frac{|00\rangle\langle00| + |11\rangle\langle11|}{2}. \tag{5.26}$$

6. Instead, if Alice chooses to send the bit "1," then she applies the operator U_i to one of the components A or A' of $|\Phi\rangle_{AA'}$. As will be demonstrated next, the whole received state by Charlie is

$$\mathcal{E}_{2,1}^{\otimes2}(U_{A \text{ or } A'}|\Phi\rangle_{AA'} \otimes |\Phi\rangle_{BB'}) = \frac{|01\rangle\langle01| + |10\rangle\langle10|}{2}. \tag{5.27}$$

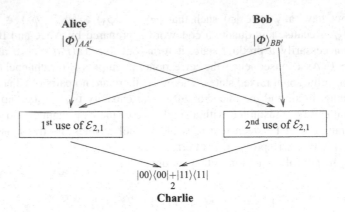

Fig. 5.9 Example of a $\mathcal{E}_{2,1}$ channel whose one-shot zero-error capacity is equal to zero. Figure illustrates how Alice and Bob can use the channel twice in order to transmit one bit to Charlie without confusion. The zero-error capacity of this channel is at least 0.5 bits/use

It is straightforward to see that the two possible output quantum states (5.26) and (5.27) are orthogonal, i.e., they can be fully distinguished by Charlie using a projective measurement. Figure 5.9 illustrates the communication protocol.

By linearity, the output of the quantum channel (5.23) after two channel uses can be written as

$$\mathcal{E}_{2,1}^{\otimes 2}(\rho) = \mathrm{Tr}\left((P_0 \otimes P_0)\rho\right)|00\rangle\langle 00| + \mathrm{Tr}\left((P_0 \otimes P_1)\rho\right)|01\rangle\langle 01|$$

$$+\mathrm{Tr}\left((P_1 \otimes P_0)\rho\right)|10\rangle\langle 10| + \mathrm{Tr}\left((P_1 \otimes P_1)\rho\right)|11\rangle\langle 11|. \quad (5.28)$$

Before the first transmission and supposing that Alice wishes to send the message "0," the global system state at the channel input is given by

$$|\Phi\rangle_{AA'} \otimes |\Phi\rangle_{BB'} = 1/4(|0000\rangle + |0011\rangle + |0022\rangle + |0033\rangle$$

$$+|1100\rangle + |1111\rangle + |1122\rangle + |1133\rangle$$

$$+|2200\rangle + |2211\rangle + |2222\rangle + |2233\rangle$$

$$+|3300\rangle + |3311\rangle + |3322\rangle + |3333\rangle). \quad (5.29)$$

Unfortunately, we cannot directly apply the above state to the composite channel (5.28) in order to get the output state after the second channel use. To see this, first note that we have written in bold the components of the global input state (5.29) that belongs to Bob. However, the expression of $\mathcal{E}_{2,1}^{\otimes 2}(\rho)$ presumes that the components of ρ must be organized in an order compatible with the original transmission protocol. For example, the second trace operator in (5.28), $\mathrm{Tr}\left((P_0 \otimes P_1)\rho\right)$, means that we must apply P_0 to the components A and B of $|\Phi\rangle_{AA'}$ and $|\Phi\rangle_{BB'}$, respectively. Analogously, the projector P_1 must be applied to the parts A' and B' of the corresponding quantum systems. For instance, we can manipulate (5.29) to fulfill this requirement. We call $\rho = |\Phi_0\rangle\langle\Phi_0|$ the state of the system corresponding to the message "0," where

$$|\Phi_0\rangle = |\Phi_{ABA'B'}\rangle$$

$$= 1/4(|0000\rangle + |0101\rangle + |0202\rangle + |0303\rangle$$

$$+|1010\rangle + |1111\rangle + |1212\rangle + |1313\rangle$$

$$+|2020\rangle + |2121\rangle + |2222\rangle + |2323\rangle$$

$$+|3030\rangle + |3131\rangle + |3232\rangle + |3333\rangle). \tag{5.30}$$

Now supposing that Alice wants to transmit the message "1" and that she applies the operator U_i to any of the components of $|\Phi\rangle_{AA'}$, we have

$$U_i|\Phi\rangle_{AA'} = (|0\rangle\langle0| - |1\rangle\langle1| + |2\rangle\langle2| - |3\rangle\langle3|)\frac{1}{2}(|00\rangle + |11\rangle + |22\rangle + |33\rangle)$$

$$= \frac{1}{2}(|00\rangle - |11\rangle + |22\rangle - |33\rangle).$$

The whole state of the system before the transmission is

$$U_i|\Phi\rangle_{AA'} \otimes |\Phi\rangle_{BB'} = 1/4(|0000\rangle + |0011\rangle + |0022\rangle + |0033\rangle$$

$$-|1100\rangle - |1111\rangle - |1122\rangle - |1133\rangle$$

$$+|2200\rangle + |2211\rangle + |2222\rangle + |2233\rangle$$

$$+|3300\rangle - |3311\rangle - |3322\rangle - |3333\rangle). \tag{5.31}$$

In the same way, we can manipulate the above state in order to apply the channel $\mathcal{E}_{2,1}(\rho)$ twice:

$$|\Phi_1\rangle = |\Phi_{ABA'B'}\rangle$$

$$= 1/4(|0000\rangle + |0101\rangle + |0202\rangle + |0303\rangle$$

$$-|1010\rangle - |1111\rangle - |1212\rangle - |1313\rangle$$

$$+|2020\rangle + |2121\rangle + |2222\rangle + |2323\rangle$$

$$-|3030\rangle - |3131\rangle - |3232\rangle - |3333\rangle). \tag{5.32}$$

Finally, the reader can verify that

$$\mathcal{E}_{2,1}^{\otimes2}(|\Phi_0\rangle\langle\Phi_0|) = \frac{|00\rangle\langle00| + |11\rangle\langle11|}{2}$$

and

$$\mathcal{E}_{2,1}^{\otimes 2}(|\Phi_1\rangle\langle\Phi_1|) = \frac{|01\rangle\langle01| + |10\rangle\langle10|}{2}$$

are orthonormal quantum states. In summary, one use of the quantum channel always leads to non-orthogonal mixed states and, therefore, the channel one-shot zero-error capacity vanishes. In contrast, using the channel twice, Alice and Bob can agree on a LOCC protocol in order to send one message to Charlie without confusion. Therefore, the asymptotical zero-error capacity of this channel is

$$C^{(0)}(\mathcal{E}_{2,1}) \geq \frac{1}{2}\log 2 = 0.5 \text{ bits per channel use.}$$

Besides its amazing feature of having the zero-error capacity activated, the channel $\mathcal{E}_{2,1}(\rho)$ has another interesting property. Alice and Bob can use the channel twice to send one bit of information to Charlie without leaking any information about the transmitted message to the other sender.

For instance, the activation of the zero-error capacity was shown in a context of a multiuser quantum channel, where senders and receivers share a classical channel to run a LOCC protocol. A natural question is whether there exist one-sender one-receiver quantum channels such that the zero-error capacity can be activated, i.e., quantum channels that a single sender cannot perfectly transmit a message to a single receiver just using the channel once, whereas such transmission is possible using the channel twice. Surprisingly, these quantum channels exist; this feature was discovered simultaneously by Duan [7] and by Cubitt et al. [5]. The so-called superactivation of the zero-error capacity is explained in the next section.

5.5.2 Superactivation of the Classical Zero-Error Capacity of Quantum Channels: Part I

In this section, we present the first of two mathematical developments that lead to a surprising result about the zero-error capacity of quantum channels. By using different frameworks, Duan [7] and Cubitt et al. [5] were able to construct families of one-sender one-receiver quantum channels whose zero-error capacities can be superactivated.

Initially, we describe the construction of two quantum channels, $\mathcal{S}$ and $\mathcal{F}$, that have a vanishing zero-error capacity. In contrast, when used together, the quantum channel $\mathcal{S} \otimes \mathcal{F}$ has a positive zero-error capacity, i.e., $C^{(0)}(\mathcal{S} \otimes \mathcal{F}) > 0$. This is not yet an example of superactivation. However, if we construct a quantum channel $\mathcal{E} = \mathcal{S} \oplus \mathcal{F}$ as the direct sum of $\mathcal{S}$ and $\mathcal{F}$,

$$\mathcal{E}(\rho) = \mathcal{S}(P_0\rho P_0) + \mathcal{F}(P_1\rho P_1), \tag{5.33}$$

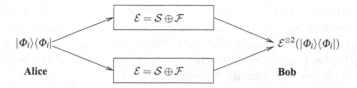

Fig. 5.10 A quantum channel $\mathcal{E}$ whose zero-error capacity can be superactivated. The channel $\mathcal{E}$ is the direct sum of $\mathcal{S}$ and $\mathcal{F}$, two quantum channels with a vanishing one-shot zero-error capacity with the property that $C^{(0)}(\mathcal{S} \otimes \mathcal{F}) > 0$. The channel $\mathcal{E}$ has a one-shot zero-error capacity equal to zero, but when used twice, Bob can perfectly distinguish between two output states $\mathcal{E}^{\otimes 2}(|\Phi_0\rangle\langle\Phi_0|)$ and $\mathcal{E}^{\otimes 2}(|\Phi_1\rangle\langle\Phi_1|)$

where P_0 and P_1 are specific projectors over the input state space, then it can be showed that the channel $\mathcal{E}$ has the one-shot zero-error capacity equal to zero, whereas classical information can be transmitted making use of the direct sum channel twice. The setup is showed in Fig. 5.10.

Consider a quantum channel $\mathcal{E}$ with Kraus operators $\mathcal{E} \equiv \{E_k\}_{k=1}^n$, where $\mathcal{E}(\rho) = \sum_k E_k \rho E_k^\dagger$ and $\sum_k E_k^\dagger E_k = I$. According to (5.8), if the quantum channel $\mathcal{E}$ has positive zero-error capacity, then there exist at least two orthogonal input states $|\psi_0\rangle, |\psi_1\rangle$ such that

$$\mathrm{Tr}\left(\mathcal{E}^\dagger(|\psi_0\rangle\langle\psi_0|)\mathcal{E}(|\psi_1\rangle\langle\psi_1|)\right) = \sum_{a,b} ||\langle\psi_0|E_a^\dagger E_b|\psi_1\rangle||^2$$

$$= 0, \tag{5.34}$$

which means that

$$\mathrm{Tr}\left(E_a^\dagger E_b|\psi_0\rangle\langle\psi_1|\right) = 0 \tag{5.35}$$

for all $1 \leq a, b \leq n$. It is evident that operators $E_a^\dagger E_b$ play an important role in studying the zero-error capacity of quantum channels. Define

$$\mathcal{K}(\mathcal{E}) = \mathrm{span}\{E_a^\dagger E_b : 1 \leq a, b \leq n\}. \tag{5.36}$$

In linear algebra, a basis $\mathcal{B}$ of a matrix space vector is called *unextendible* if $\mathcal{B}^\perp$ contains no rank-one matrices. Consequently, $\mathcal{B}^\perp$ contains only matrices with rank two or more and, therefore, we say that $\mathcal{B}^\perp$ is a completely entangled state space. When $\mathcal{B}^\perp$ contains at least one rank-one matrix, the basis $\mathcal{B}$ is called *extendible*. This kind of partition of a state space has interesting applications in quantum information theory, specially in distinguishability of general quantum states and subspaces. Some references about unextendible basis can be found at Further Reading section. For our purposes, we just mention an important property of unextendible basis. It was shown that if the dimension of a matrix subspace $\mathcal{B}$ is $\dim(\mathcal{B}) < 2d - 1$, where d is the dimension of the whole matrix space, then $\mathcal{B}$ is always extendible.

We turn our attention to (5.35). First note that the operator $|\psi_0\rangle\langle\psi_1|$ is orthogonal to $E_a^\dagger E_b$ for all $1 \leq a,b \leq n$, i.e., $|\psi_0\rangle\langle\psi_1| \in \mathcal{K}(\mathcal{E})^\perp$. Moreover, the operator $|\psi_0\rangle\langle\psi_1|$ has rank equal to one. Therefore, we conclude that if a quantum channel $\mathcal{E} \equiv \{E_k\}_{k=1}^n$ has positive one-shot zero-error capacity, then $\mathcal{K}(\mathcal{E})$ is extendible. The converse is also true, as states the following lemma [7].

Lemma 5.1. *Let $\mathcal{E} \equiv \{E_k\}_{k=1}^n$ be a quantum channel with Kraus operators E_k. The channel $\mathcal{E}$ has positive zero-error capacity if and only if $\mathcal{K}(\mathcal{E})$ is extendible.*

Another property of $\mathcal{K}(\mathcal{E})$ is $\mathcal{K}(\mathcal{E})^\dagger = \mathcal{K}(\mathcal{E})$. Note that $\mathcal{K}(\mathcal{E})^\dagger = \{K^\dagger, K \in \mathcal{K}(\mathcal{E})\}$. Moreover, because $\mathcal{E}$ is trace preserving, it turns out that $I \in \mathcal{K}(\mathcal{E})$. In fact, these two properties are necessary and sufficient conditions to guarantee the existence of a quantum channel from an input state space of operators $\mathcal{B}(\mathcal{H}_d)$ to the output state space $\mathcal{B}(\mathcal{H}_{d'})$.

Lemma 5.2 (Duan [7]). *Let $\mathcal{M}$ be a matrix subspace of $\mathcal{B}(\mathcal{H}_d)$. There is a quantum channel $\mathcal{E}$ from $\mathcal{B}(\mathcal{H}_d)$ to $\mathcal{B}(\mathcal{H}_{d'})$ such that $\mathcal{K}(\mathcal{E}) = \mathcal{M}$ if and only if $\mathcal{M}^\dagger = \mathcal{M}$ and $I \in \mathcal{M}$.*

The conditions in Lemmas 5.1 and 5.2 are important because they allow to construct quantum channels whose zero-error capacity can be superactivated. As already mentioned, this can be achieved by finding two quantum channels, $\mathcal{S}$ and $\mathcal{F}$, with vanish zero-error capacity, whereas $\mathcal{S} \otimes \mathcal{F}$ has positive zero-error capacity. This can be done by writing down two partitions of a state space, say $\mathcal{K}(\mathcal{S})$ and $\mathcal{K}(\mathcal{F})$, with the following properties: $\mathcal{K}(\mathcal{S})$ and $\mathcal{K}(\mathcal{F})$ are unextendible, whereas $\mathcal{K}(\mathcal{S}) \otimes \mathcal{K}(\mathcal{F})$ is extendible.

Example 5.6 (Superactivation of the Zero-Error Capacity of Quantum Channels).
Let $\mathcal{K}(\mathcal{S})$ be the matrix state space spanned by the vectors:

$$S_1 = |0\rangle\langle0| + |1\rangle\langle1|,$$
$$S_2 = |2\rangle\langle2| + |3\rangle\langle3|,$$
$$S_3 = |2\rangle\langle0| - |0\rangle\langle2|,$$
$$S_4 = |3\rangle\langle0| + |0\rangle\langle3|,$$
$$S_5 = |1\rangle\langle3| + |3\rangle\langle1|,$$
$$S_6 = \cos\theta|0\rangle\langle1| + \sin\theta|2\rangle\langle3| - |1\rangle\langle2|,$$
$$S_7 = \cos\theta|1\rangle\langle0| + \sin\theta|3\rangle\langle2| - |2\rangle\langle1|,$$
$$S_8 = \sin\theta|0\rangle\langle1| - \cos\theta|2\rangle\langle3| - \sin\theta|1\rangle\langle0| + \cos\theta|3\rangle\langle2|,$$

$$(5.37)$$

where $0 < \theta < \pi/2$ is a parameter. The reader can verify that $\mathcal{K}(\mathcal{S})^\perp = \mathcal{K}(\mathcal{S})$ and $I \in \mathcal{K}(\mathcal{S})$. In addition, consider the matrix state space spanned by the following vectors:

$$F_1 = |0\rangle\langle 0| - |1\rangle\langle 1|,$$

$$F_2 = |2\rangle\langle 2| - |3\rangle\langle 3|,$$

$$F_3 = |2\rangle\langle 0| + |0\rangle\langle 2|,$$

$$F_4 = |3\rangle\langle 0| - |0\rangle\langle 3|, \tag{5.38}$$

$$F_5 = |1\rangle\langle 3| - |3\rangle\langle 1|,$$

$$F_6 = \cos\theta|0\rangle\langle 1| + \sin\theta|2\rangle\langle 3| + |1\rangle\langle 2|,$$

$$F_7 = \cos\theta|1\rangle\langle 0| + \sin\theta|3\rangle\langle 2| + |2\rangle\langle 1|,$$

$$F_8 = -\sin\theta|0\rangle\langle 1| + \cos\theta|2\rangle\langle 3| + \sin\theta|1\rangle\langle 0| - \cos\theta|3\rangle\langle 2|.$$

The reader can easily verify that $\mathcal{K}(\mathcal{F})$ satisfies the Hermitian condition, $\mathcal{K}(\mathcal{F})^\dagger = \mathcal{K}(\mathcal{F})$. Moreover, the subspaces $\mathcal{K}(\mathcal{S})$ and $\mathcal{K}(\mathcal{F})$ are orthogonal with respect to the Hilbert-Schmidt inner product.

The two matrix vectors state spaces $\mathcal{K}(\mathcal{S})$ and $\mathcal{K}(\mathcal{F})$ has the following desirable properties:

(a) $\mathcal{K}(\mathcal{S})$ and $\mathcal{K}(\mathcal{F})$ are unextendible, i.e., they are completely entangled state spaces;
(b) $\mathcal{K}(\mathcal{S}) \otimes \mathcal{K}(\mathcal{F})$ is extendible.

In order to verify property (a), define a rank-one matrix $|\psi\rangle\langle\phi|$ orthogonal to $\mathcal{K}(\mathcal{S})$, where $|\psi\rangle = \sum_{i=0}^{3} c_i|i\rangle$ and $|\phi\rangle = \sum_{j=0}^{3} d_j|j\rangle$. Then, for each $k = 1, \ldots, 8$, $\mathrm{Tr}\,(S_k|\psi\rangle\langle\phi|) - 0$ implies $c_i = d_i = 0$ for all $0 \leq i \leq 3$, i.e., the orthogonal complement of $\mathcal{K}(\mathcal{S})$ has no rank-one matrices and, therefore, $\mathcal{K}(\mathcal{S})$ is unextendible. The same holds for the subspace $\mathcal{K}(\mathcal{F})$.

Property (b) can be demonstrated by defining the quantum state

$$|\Phi_0\rangle = \frac{|00\rangle + |11\rangle + |22\rangle + |33\rangle}{2} \tag{5.39}$$

and the operator $U = |0\rangle\langle 0| - |1\rangle\langle 1| + |2\rangle\langle 2| - |3\rangle\langle 3|$. The reader can verify that the following quantum state

$$|\Phi_1\rangle = (I \otimes U)\frac{|00\rangle + |11\rangle + |22\rangle + |33\rangle}{2} \tag{5.40}$$

gives rise to the rank-one matrix $|\Phi_1\rangle\langle\Phi_1|$ orthogonal to $\mathcal{K}(\mathcal{S}) \otimes \mathcal{K}(\mathcal{S})$, i.e.,

$$\mathrm{Tr}\,(S_i \otimes F_j|\Phi_1\rangle\langle\Phi_1|) = 0 \qquad \forall i, j = 1, \ldots, 8.$$

Therefore, the vector space $\mathcal{K}(\mathcal{S}) \otimes \mathcal{K}(\mathcal{S})$ is extendible because $(\mathcal{K}(\mathcal{S}) \otimes \mathcal{K}(\mathcal{S}))^\perp$ has a rank-one matrix.

According to Lemma 5.2, the corresponding quantum channels $\mathcal{S}$ and $\mathcal{F}$ have no zero-error capacity when used individually, whereas the channel $\mathcal{S} \otimes \mathcal{F}$ has positive zero-error capacity. By using $S \otimes F$, a sender can prepare one of the entangled states $|\Phi_0\rangle$ and $|\Phi_1\rangle$ to transmit a classical message to a receiver, since the latter can perfectly distinguish between $S \otimes F(|\Phi_0\rangle)$ and $S \otimes F(|\Phi_1\rangle)$.

As already mentioned, this is not yet an example of superactivation, since S and F are different channels. However, if we consider the direct sum channel $S \oplus F$ (5.33),

$$\mathcal{E}(\rho) = S(P_0 \rho P_0) + \mathcal{F}(P_1 \rho P_1),$$

with $P_0 = |\Phi_0\rangle\langle\Phi_0|$ and $P_1 = I - |\Phi_0\rangle\langle\Phi_0|$, then it is clear that a single use of the channel $\mathcal{E}$ cannot transmit classical information without error. In contrast, one can verify that when the channel $\mathcal{E}$ is used twice, then Bob is able to distinguish between the two orthogonal states, $\mathcal{E}^{\otimes 2}(|\Phi_0\rangle)$ and $\mathcal{E}^{\otimes 2}(|\Phi_1\rangle)$. Therefore, the use of entanglement between two uses of a quantum channel can superactivate the zero-error capacity of the channel. Finally, we can conclude that the asymptotic zero-error capacity of $\mathcal{E}$ is

$$C^{(0)}(\mathcal{E}) \geq 0.5 \text{ bits per channel use.}$$

A short remark on the construction of the quantum channels S and F should be given. First, note that

$$M = \sum_{i=1}^{8} S_i^\dagger S_i = \sum_{i=1}^{8} F_i^\dagger F_i,$$

where

$$M = \begin{bmatrix} 4 & 0 & 0 & 0 \\ 0 & 4 & 0 & 0 \\ 0 & 0 & 4 & 0 \\ 0 & 0 & 0 & 4 \end{bmatrix}.$$

In order to construct the corresponding trace-preserving quantum operations $\mathcal{E}$ we only need to define the sets

$$S \equiv \left\{ S_i M^{-\frac{1}{2}} \right\}_{i=1}^{8} \quad \text{and} \quad F \equiv \left\{ F_i M^{-\frac{1}{2}} \right\}_{i=1}^{8}, \tag{5.41}$$

where

$$M^{-\frac{1}{2}} = \begin{bmatrix} 0.5 & 0 & 0 & 0 \\ 0 & 0.5 & 0 & 0 \\ 0 & 0 & 0.5 & 0 \\ 0 & 0 & 0 & 0.5 \end{bmatrix}.$$

5.5.3 Superactivation of the Classical Zero-Error Capacity of Quantum Channels: Part II

This section describes a different approach to determine necessary and sufficient conditions for the existence of quantum channels, S and $\mathcal{F}$, whose one-shot zero-error capacities are zero, while the composite channel $S \otimes \mathcal{F}$ has positive (one-shot) zero-error capacity.

Consider a quantum channel $S : \mathcal{B}(\mathcal{H}) \to \mathcal{B}(\mathcal{H})$. The channel S has a vanishing one-shot zero-error capacity if and only if all quantum states in $\mathcal{H}$ are adjacent, i.e.,

$$\forall |\psi\rangle, |\varphi\rangle \in \mathcal{H} : \mathrm{Tr}\left(S(|\psi\rangle)^{\dagger}S(|\varphi\rangle)\right) \neq 0. \tag{5.42}$$

Let S^* be the adjoint[2] (or dual) of S with respect to the Hilbert-Schmidt inner product. Because the cyclic property of the trace,

$$\mathrm{Tr}\left(S(|\psi\rangle)^{\dagger}S(|\varphi\rangle)\right) = \mathrm{Tr}\left(\psi \cdot S(S(|\varphi\rangle))\right)$$
$$= \mathrm{Tr}\left(\psi \cdot S^* \circ S(|\varphi\rangle)\right), \tag{5.43}$$

where $\psi \equiv |\psi\rangle\langle\psi|$ and $S(S(\cdot)) \equiv S \circ S(\cdot)$ were defined for short.

Conversely, a quantum channel has positive zero-error capacity if and only if there exists at least one pair of non-adjacent states in $\mathcal{H}$, i.e.,

$$\exists |\psi\rangle, |\varphi\rangle \in \mathcal{H} : \mathrm{Tr}\left(S(|\psi\rangle)^{\dagger}S(|\varphi\rangle)\right) = 0, \tag{5.44}$$

or, equivalently,

$$\exists |\psi\rangle, |\varphi\rangle \in \mathcal{H} : \mathrm{Tr}\left(\psi \cdot S^* \circ S(|\varphi\rangle)\right) = 0. \tag{5.45}$$

The problem of finding quantum channels whose zero-error capacities can be superactivated is reformulated as follows. One needs to find two quantum channels $S, \mathcal{F}$, such that

(a)

$$\forall |\psi\rangle, |\varphi\rangle \in \mathcal{H} : \mathrm{Tr}\left(\psi \cdot S^* \circ S(|\varphi\rangle)\right) \neq 0, \tag{5.46}$$

which means $C^{(0)}(S) = 0$;

---

[2] The adjoint of $S : \mathcal{B}(\mathcal{H}) \to \mathcal{B}(\mathcal{H})$ is dual with respect to the Hilbert-Schmidt inner product such that $\mathrm{Tr}\left(\rho^{\dagger}S(\sigma)\right) = \mathrm{Tr}\left(S^*(\rho)^{\dagger}\sigma\right)$, $\rho, \sigma \in \mathcal{B}(\mathcal{H})$. If $S \equiv \{E_k\}$ is defined by a set of Kraus operator E_k, then $S^* \equiv \{E_k^{\dagger}\}$.

(b)

$$\forall |\psi\rangle, |\varphi\rangle \in \mathcal{H} : \mathrm{Tr}\left(\psi \cdot \mathcal{F}^* \circ \mathcal{F}(|\varphi\rangle)\right) \neq 0, \qquad (5.47)$$

which means $C^{(0)}(\mathcal{F}) = 0$;

(c)

$$\exists |\alpha\rangle, |\phi\rangle \in \mathcal{H}^{\otimes 2} : \mathrm{Tr}\left(\alpha \cdot (\mathcal{S}^* \circ \mathcal{S}) \otimes (\mathcal{F}^* \circ \mathcal{F})(|\phi\rangle)\right) = 0, \qquad (5.48)$$

which means $C^{(0)}(\mathcal{S} \otimes \mathcal{F}) > 0$.

The composite map $\mathcal{S}^* \circ \mathcal{S}$ plays an important role in studying the zero-error capacity. A map $\mathcal{N}$ is called *conjugate-divisible* if it can be decomposed as $\mathcal{N} = \mathcal{S}^* \circ \mathcal{S}$. Before enunciating a theorem that establishes necessary and sufficient conditions for the existence of conjugate-divisible maps and gives a full characterization of its corresponding Choi-Jamiołkowski matrices, it is helpful to define positive-semidefinite states and subspaces, as well as conjugate-symmetric states and subspaces.

For bipartite states $|\phi\rangle_{AB} \in \mathcal{H}_{AB}$ with basis $|i_A\rangle |j_B\rangle$, there exists an isomorphism with the space of $d_A \times d_A$ matrices in the following way:

$$|\phi\rangle_{AB} = \sum_{ij} M_{ij} |i_A\rangle |j_B\rangle.$$

In this way, the bipartite state $|\phi\rangle_{AB}$ is said to be positive-semidefinite if the corresponding matrix $M_{|\phi\rangle_{AB}} = [M_{ij}]$ is positive-semidefinite. Analogously, a subspace $\mathcal{H}_{AB}$ is positive-semidefinite if it can be spanned by a set of positive-semidefinite states.

A bipartite state or operator $|\phi\rangle_{AB} \in \mathcal{H}_{AB}$ is conjugate-symmetric in a given basis $|i_A\rangle |j_B\rangle$ if it is invariant under the *flip* operation:

$$\mathbb{F}\left(\sum_{ij} c_{ij} |i_A\rangle |j_B\rangle\right) = \sum_{ij} \bar{c}_{ij} |j_A\rangle |i_B\rangle.$$

The effect of the flip operation is to interchange the two parties and complex conjugation. Similarly, a subspace is said to be conjugate-symmetric if it is invariant under the same operation.

Theorem 5.3 (Existence of Conjugate-Divisible Maps [5]). *Given a subspace* $\mathcal{H}_{AB} \subseteq \mathcal{H}^{\otimes 2}$ *such that supp* $(Tr_B (\mathcal{H}_{AB})) = \mathcal{H}$, *there exists a conjugate-divisible map with (in general non-standard) Choi-Jamiołkowski matrix* σ_{AB} *such that supp* $(\sigma_{AB}) = \mathcal{H}_{AB}$ *if and only if* $\mathcal{H}_{AB}$ *is a positive-semidefinite and conjugate-symmetric subspace.*

The notation supp $(\mathrm{Tr}_B(\mathcal{H}_{AB}))$ stands for $\bigcup_{|\phi\rangle \in \mathcal{H}_{AB}}$ supp $(\mathrm{Tr}_B(|\phi\rangle\langle\phi|))$. Now requirements (a) to (c) can be converted into necessary and sufficient conditions for the maps to satisfy (5.46) to (5.48). Let σ_S and $\sigma_{\mathcal{F}}$ be the Choi-Jamiołkowski matrices corresponding to the conjugate-divisible maps $\mathcal{N}_S = \mathcal{S}^* \circ \mathcal{S}$ and $\mathcal{N}_{\mathcal{F}} = \mathcal{F}^* \circ \mathcal{F}$, respectively. Equations (5.46) and (5.47) can be rewritten as

$$\forall |\psi\rangle, |\varphi\rangle \in \mathcal{H} : \mathrm{Tr}\left(\psi \cdot \mathcal{S}^* \circ \mathcal{S}(|\varphi\rangle)\right) = \mathrm{Tr}\left(\psi \cdot \mathrm{Tr}_A\left(\sigma_S \cdot \varphi^T \otimes \mathbb{1}\right)\right)$$
$$= \mathrm{Tr}\left(\sigma_S \cdot \varphi \otimes \psi\right)$$
$$\neq 0, \tag{5.49}$$

and

$$\forall |\psi\rangle, |\varphi\rangle \in \mathcal{H} : \mathrm{Tr}\left(\sigma_{\mathcal{F}} \cdot \varphi \otimes \psi\right) \neq 0. \tag{5.50}$$

Therefore, if $\mathcal{H}_{\sigma_S}, \mathcal{H}_{\sigma_{\mathcal{F}}} \subseteq \mathcal{H}^{\otimes 2}$ are the subspaces spanned by the support of σ_S and $\sigma_{\mathcal{F}}$, respectively, it is necessary and sufficient to require that their orthogonal complements contain no product states, i.e.,

$$\nexists |\psi\rangle, |\varphi\rangle \in \mathcal{H} : |\psi\rangle \otimes |\varphi\rangle \in \mathcal{H}_{\sigma_S}^\perp, \mathcal{H}_{\sigma_{\mathcal{F}}}^\perp. \tag{5.51}$$

We turn our attention to (5.48) in order to find necessary and sufficient conditions to the joint map $\mathcal{S} \otimes \mathcal{F}$ to fulfill the corresponding requirement. Without loss of generality, fix the states $|\alpha\rangle$ and $|\phi\rangle$ to be maximally entangled in the following way. Let $|\omega\rangle$ be the full rank (unnormalized) state $|\omega\rangle = \sum_i |i\rangle|i\rangle$ and define $|\alpha\rangle = (U \otimes V)|\omega\rangle$, $|\phi\rangle = (W \otimes X)|\omega\rangle$, where U, V, W, X are unitary. Again, if σ_S and $\sigma_{\mathcal{F}}$ are the Choi-Jamiołkowski matrices corresponding to the conjugate-divisible maps $\mathcal{N}_S = \mathcal{S}^* \circ \mathcal{S}$ and $\mathcal{N}_{\mathcal{F}} = \mathcal{F}^* \circ \mathcal{F}$, respectively, then

$$0 = \mathrm{Tr}\left(\alpha \cdot (\mathcal{S}^* \circ \mathcal{S}) \otimes (\mathcal{F}^* \circ \mathcal{F})(|\phi\rangle)\right)$$
$$= \mathrm{Tr}\left(\alpha \cdot \mathcal{N}_S \otimes \mathcal{N}_{\mathcal{F}}(|\phi\rangle)\right)$$
$$= \mathrm{Tr}\left(\alpha \cdot \mathrm{Tr}_A\left(\sigma_S \otimes \sigma_F \cdot \phi^T \otimes \mathbb{1}\right)\right)$$
$$= \mathrm{Tr}\left(\sigma_S \otimes \sigma_F \cdot \phi^T \otimes \alpha\right)$$
$$= \mathrm{Tr}\left(\sigma_S \otimes \sigma_F \cdot (U \otimes V\omega^T U^T \otimes V^T) \otimes (W \otimes X\omega W^\dagger \otimes X^\dagger)\right)$$
$$= \mathrm{Tr}\left(\left(U \otimes W\sigma_S U^T \otimes W^\dagger\right)^T \cdot \left(V \otimes X\sigma_F V^T \otimes X^\dagger\right)\right)$$
$$= \mathrm{Tr}\left(\sigma_S^T \cdot (U \otimes V\sigma_{\mathcal{F}} U^\dagger \otimes V^\dagger)\right). \tag{5.52}$$

Besides fulfilling the requirements (5.51), the support of the Choi-Jamiołkowski matrices σ_S and $\sigma_{\mathcal{F}}$ must be related by

$$\mathcal{H}_{\sigma_{\mathcal{F}}}^T = (U \otimes V)\mathcal{H}_{\sigma_S}^\perp.$$

Because conjugate-symmetry, Schmidt-rank, and positive-semidefiniteness are invariant under the transpose operation, we can write

$$\mathcal{H}_{\sigma_{\mathcal{F}}} = (U \otimes V)\mathcal{H}_{\sigma_S}^{\perp}. \tag{5.53}$$

Moreover, if a subspace is conjugate-symmetric, then so is its orthogonal complement. Considering the fact that Schmidt-rank is invariant under unitary operations, all the conditions can be expressed in terms of a single subspace $\mathcal{H}_2 \subseteq \mathcal{H}^{\otimes 2}$. Finally, (5.51) and (5.53), together with Theorem 5.3, give necessary and sufficient conditions to guarantee the existence of two quantum channels S and $\mathcal{F}$ such that $C^{(0)}(\mathcal{E}) = C^{(0)}(\mathcal{F}) = 0$, while the composite channel $S \otimes \mathcal{F}$ has positive one-shot zero-error capacity. All of these conditions are grouped in Theorem 5.4

Theorem 5.4 (Superactivation of the One-Shot Zero-Error Capacity [5]). *If there exists a subspace $\mathcal{H}_2 \subseteq \mathcal{H}^{\otimes 2}$ and unitaries U, V satisfying*

$$\nexists |\psi\rangle, |\varphi\rangle \in \mathcal{H} \; : \; |\psi\rangle \otimes |\varphi\rangle \in \mathcal{H}_2^{\perp}, \tag{5.54}$$

$$\nexists |\psi\rangle, |\varphi\rangle \in \mathcal{H} \; : \; |\psi\rangle \otimes |\varphi\rangle \in \mathcal{H}_2, \tag{5.55}$$

$$\mathbb{F}(\mathcal{H}_2) = \mathcal{H}_2, \tag{5.56}$$

$$\mathbb{F}(U \otimes V \cdot \mathcal{H}_2) = U \otimes V \cdot \mathcal{H}_2, \tag{5.57}$$

$$\exists \{M_i \geq 0\} \; : \; \mathcal{H}_2 \equiv span\{M_i\}, \tag{5.58}$$

$$\exists \{M_j \geq 0\} \; : \; U \otimes V \cdot \mathcal{H}_2^T \equiv span\{M_j\}, \tag{5.59}$$

then there exist quantum channels S and $\mathcal{F}$ whose one-shot zero-error capacity is zero, whereas the joint channel $S \otimes \mathcal{F}$ has positive zero-error capacity.

In Theorems 5.4, (5.54) and (5.55) fulfill requirements (5.46) and (5.47), i.e., they impose that the one-shot zero-error capacity of the channels S and $\mathcal{F}$ are both equal to zero. Equation (5.57) ensures that the joint channel $S \otimes \mathcal{F}$ has positive zero-error capacity, whereas (5.56), (5.58), and (5.59) are necessary and sufficient conditions to guarantee the existence of the corresponding quantum channels, as stated in Theorem 5.3.

5.6 Further Reading

In this chapter, we revisited the classical zero-error capacity of quantum channels proposed by Medeiros [11]. This chapter is based on his thesis, but many articles published previously built up his theory [12–14, 16].

Many alternative definitions to the zero-error capacity of a quantum channel can also be found in the literature. Medeiros and Assis proposed a version in which the maximum amount of quantum information sent through quantum channels without errors is considered, the so-called quantum zero-error capacity of a quantum channel [14]. Other variants proposed by Winter et al. are shown in Chap. 8 and can also be found detailed in papers by these authors [4, 6, 9].

Superactivation was first described by Duan and Shi [8] for a scenario of multiple senders and receivers. Using the concept of unextendible basis [3, 19, 22], Duan [7] demonstrated the existence of one-sender one-receiver quantum channel whose zero-error capacity can be activated. This phenomena was independently studied by Cubitt et al., which proved a more general result on the superactivation of the asymptotic zero-error capacity [5]. Park and Lee [18] showed that the zero-error capacity of qubit channels cannot be superactivated.

Cubitt and Smith [2] considered the scenario where two quantum channels $\mathcal{S}$ and $\mathcal{F}$ have a vanishing zero-error capacity, whereas the joint channel $\mathcal{S} \otimes \mathcal{F}$ could transmit quantum information at a positive rate and with probability of error equal to zero. The authors called this effect the super-duper-activation of the quantum zero-error capacity. Various examples of low dimensional quantum channels whose one-shot classical and quantum zero-error capacities can be superactivated were described by Shirokov and Shulman [21].

References

1. Cover TM, Thomas JA (1991) Elements of information theory. Wiley, New York
2. Cubitt TS, Smith G (2012) An extreme form of superactivation for quantum zero-error capacities. IEEE Trans Inf Theory 58(3):1953–1961
3. Cubitt T, Harrow AW, Leung D, Montanaro A, Winter A (2008) Counterexamples to additivity of minimum output p-rényi entropy for p close to 0. Commun Math Phys 284(1):281–290. doi:10.1007/s00220-008-0625-z, http://dx.doi.org/10.1007/s00220-008-0625-z
4. Cubitt TS, Leung D, Matthews W, Winter A (2010) Improving zero-error classical communication with entanglement. Phys Rev Lett 104:230503. doi:http://dx.doi.org/10.1103/PhysRevLett.104.230503
5. Cubitt TS, Chen J, Harrow AW (2011) Superactivation of the asymptotic zero-error classical capacity of a quantum channel. IEEE Trans Inf Theory 57(12):8114–8126. doi:10.1109/TIT.2011.2169109
6. Cubitt TS, Leung D, Matthews W, Winter A (2011) Zero-error channel capacity and simulation assisted by non-local correlations. IEEE Trans Inf Theory 57(8):5509–5523
7. Duan R (2009) Super-activation of zero-error capacity of noisy quantum channels. arxiv:quant-ph/0906.2527v1
8. Duan R, Shi Y (2008) Entanglement between two uses of a noisy multipartite quantum channel enables perfect transmission of classical information. Phys Rev Lett 101:020501. doi:10.1103/PhysRevLett.101.020501, http://link.aps.org/doi/10.1103/PhysRevLett.101.020501
9. Duan R, Severini S, Winter A (2011) Zero-error communication via quantum channels, non-commutative graphs and a quantum Lovasz ϑ function. In: IEEE international symposium on information theory, Russia, pp 64–68
10. Holevo AS (1998) The capacity of the quantum channel with general signal states. IEEE Trans Inf Theory 4(1):269–273
11. Medeiros RAC (2008) Zero-error capacity of quantum channels. Ph.D Thesis, Universidade Federal de Campina Grande – TELECOM Paris Tech

12. Medeiros RAC, de Assis FM (2004) Zero-error capacity of a quantum channel. Lect Notes Comput Sci 3124:100–105
13. Medeiros RAC, de Assis FM (2005) Capacidade erro-zero de canais quânticos e estados puros. In: Simpósio Brasileiro de Telecomunicações, Campinas, São Paulo, pp 1–6
14. Medeiros RAC, de Assis FM (2005) Quantum zero-error capacity. Int J Quantum Inf 3(1):135–139
15. Medeiros RA, Alleaume R, Cohen G, de Assis FM (2006) Zero-error capacity of quantum channels and noiseless subsystems. In: IEEE international telecommunications symposium, Fortaleza, Brazil, pp 900–905
16. Medeiros RAC, Alleaume R, Cohen G, de Assis FM (2006) Quantum states characterization for the zero-error capacity. http://arxiv.org/abs/quant-ph/0611042. Accessed 25 Oct 2013
17. Nielsen MA, Chuang IL (2010) Quantum computation and quantum information. Cambridge University Press, Cambridge
18. Park J, Lee S (2012) Zero-error classical capacity of qubit channels cannot be superactivated. Phys Rev A 85:052321
19. Parthasarathy KR (2004) On the maximal dimension of a completely entangled subspace for finite level quantum systems. Proc Math Sci 114(4):365–374. doi:10.1007/BF02829441, http://dx.doi.org/10.1007/BF02829441
20. Schumacher B, Westmoreland MD (1997) Sending classical information via noisy quantum channels. Phys Rev A 56:131–138. doi:10.1103/PhysRevA.56.131
21. Shirokov ME, Shulman T (2015) On superactivation of zero-error capacities and reversibility of a quantum channel. Commun Math Phys 335(3):1159–1179
22. Walgate J, Scott AJ (2008) Generic local distinguishability and completely entangled subspaces. J Phys A Math Theor 41(37):375305. http://stacks.iop.org/1751-8121/41/i=37/a=375305

Chapter 6
Zero-Error Secrecy Capacity

Quantum key distribution is one of the most settled techniques nowadays to perform secure communications over quantum channels [42, p. 586]. Even though its security proofs are well established [37], in practical scenarios many of these protocols are not adequate due to noise in the quantum channel. The noise does not only increase the error rate in the transmission, but can also hinder eavesdropping detection in a process of security control [34].

Considering the practical difficulties to perform secure communications in noisy quantum channels, this chapter introduces some recent results regarding the *zero-error secrecy capacity* (ZESC), the higher transmission rate that can be achieved in certain noisy quantum channels that allows information to be sent without errors and in an unconditionally secure way. This capacity unifies concepts from quantum zero-error information theory, from quantum secrecy capacity of quantum channels, and also from decoherence-free subspaces and subsystems.

To present such developments, this chapter is organized as follows. Some background concepts of decoherence-free subspaces and subsystems are presented in Sect. 6.1. Section 6.2 discusses the quantum secrecy capacity. The model of communications and the formalism of concepts and proofs regarding ZESC are shown in Sect. 6.3. The relation between ZESC and graph theory is elucidated in Sect. 6.4. After that, the security level that this approach provides is presented in Sect. 6.5. Detailed examples considering different scenarios for the ZESC are illustrated in Sect. 6.6. Recent works in literature that have intersections with the ZESC, and that may point to further work are introduced in Sect. 6.7. Lastly, further reading is suggested in Sect. 6.8.

© Springer International Publishing Switzerland 2016
E.B. Guedes et al., *Quantum Zero-Error Information Theory*,
DOI 10.1007/978-3-319-42794-2_6

6.1   Decoherence-Free Subspaces and Subsystems

Suppose a closed quantum system composed by a system of interest, denoted by S and defined in a Hilbert space $\mathcal{H}$, and by the environment, denoted by E. This system has the following Hamiltonian:

$$\mathbb{H} = \mathbb{H}_S \otimes \mathbb{1}_E + \mathbb{1}_S \otimes \mathbb{H}_E + \mathbb{H}_{SE}, \tag{6.1}$$

where $\mathbb{1}$ is the identity operator, $\mathbb{H}_S$ denotes the operator of the system of interest, $\mathbb{H}_E$ denotes the operator of the environment, and $\mathbb{H}_{SE}$ denotes the operator of the interaction between system and environment [34].

To a complete absence of errors, the ideal scenario happens when $\mathbb{H}_{SE}$ is zero, indicating that system and environment are completely decoupled and evolved unitary according to their own Hamiltonians $\mathbb{H}_S$ and $\mathbb{H}_E$, respectively [34]. However, in realistic situations, this ideal scenario does not occur since no system can be completely free of errors. So, after isolating a system as better as possible, we must adopt at least one of the following strategies: identify and correct errors when they occur; avoid error as much as possible; suppress the error of the system [5].

If some symmetries exist in the interaction between system and environment, it is possible to find a "safe place" in the Hilbert space that is not subject to the negative effects of decoherence. Let $\{A_i(t)\}$ be a set of operators in the operator-sum representation (OSR) describing the evolution of a system. We say that a density matrix ρ_S is invariant under the operators $\{A_i(t)\}$ if $\sum_i A_i(t)\rho_S A_i^\dagger(t) = \rho_S$. Taking this into account, we can define the *decoherence-free subspaces and subsystems* (DFS) whose states are invariant despite a non-trivial coupling between system and environment.

Definition 6.1 (Decoherence-Free Subspaces and Subsystems [1]). A subspace $\tilde{\mathcal{H}}$ from a Hilbert space $\mathcal{H}$ is said to be decoherence-free regarding the coupling between system and environment if every pure state in this subspace is invariant under the OSR evolution, despite any environment initial condition, i.e.,

$$\sum_i A_i(t)|\tilde{k}\rangle\langle\tilde{k}|A_i^\dagger(t) = |\tilde{k}\rangle\langle\tilde{k}|, \forall |\tilde{k}\rangle\langle\tilde{k}| \in \tilde{\mathcal{H}}, \forall \rho_E(0). \tag{6.2}$$

Let the Hamiltonian of the interaction between system and environment be $\mathbb{H}_{SE} = \sum_j S_j \otimes E_j$, where S_j and E_j are the operators of the system and the environment, respectively. We consider that the environment operators E_j are linearly independent. The symmetries required to the existence of a DFS are described as follows, whose proof is shown in [34, Sect. 5].

Theorem 6.1 (Conditions for the Existence of Decoherence-Free Subspaces). *A subspace $\tilde{\mathcal{H}}$ is decoherence-free if and only if the system operators S_j act proportionally to the identity in this subspace, i.e.,*

$$S_j|\tilde{k}\rangle = c_j|\tilde{k}\rangle \qquad \forall j, |\tilde{k}\rangle \in \tilde{\mathcal{H}}. \tag{6.3}$$

The notion of a subspace that remains decoherence-free during the system evolution is not, however, the most general way to decoherence-free encoding in quantum systems [34]. Knill et al. [32] developed a method to encoding into subsystems instead of subspaces.

Definition 6.2 (Decoherence-Free Subsystems). Let $\mathcal{E} : \mathcal{B}(\mathcal{H}) \rightarrow \mathcal{B}(\mathcal{H})$ be a positive trace-preserving superoperator in a Hilbert space $\mathcal{H}$. Suppose $\mathcal{H} = (\mathcal{H}^A \otimes \mathcal{H}^B) \oplus \mathcal{K}$. We say that $\mathcal{H}^B$ (dim($\mathcal{H}^B$) ≥ 1) is a *decoherence-free subsystem* if, $\forall \sigma^A \in B(\mathcal{H}^A)$ and $\forall \sigma^B \in B(\mathcal{H}^B)$, there is $\tau^A \in B(\mathcal{H}^A)$ such that

$$\mathcal{E}(\sigma^A \otimes \sigma^B) = \tau^A \otimes \sigma^B. \tag{6.4}$$

We can also write this definition using the partial trace:

$$\mathrm{Tr}_A \left[\mathcal{E}(\sigma) \right] = \mathrm{Tr}_A(\sigma) \qquad \forall \sigma = \sigma^A \otimes \sigma^B. \tag{6.5}$$

When dim($\mathcal{H}^A$) $= 1$, we say that $\mathcal{H}^B$ is a decoherence-free subspace for $\mathcal{E}$.

It is possible to build codes from states of a DFS which are known as *quantum error-avoiding codes* (QEAC). Information encoded into DFS is not affected by the channel's noise. Therefore, no error-correcting procedure is necessary. Error-avoiding codes can be contrasted with *quantum error-correcting codes* (QECC) regarding some aspects: QECCs are designed to correct errors after they occur, while QEACs do not have abilities to correct errors, because they avoid it; the most adopted QECCs are non-degenerated, while QEACs are highly degenerated codes; QEACs usually require less physical qubits to represent a logical qubit when compared to QECCs. In particular, if the degenerescence of a QECC reaches the maximum, then this code is reduced to a QEAC, showing a situation where one kind of code becomes equivalent to the other [14].

Even though DFS is a way to avoid errors, not all situations attain symmetry requirements to the existence of such subspaces. Zanardi and Rasetti [58] state that such conditions occur only if there is *collective decoherence* which occurs when several qubits couple in an identical way with the environment while undergoing both dephasing and dissipation.

Example 6.1 (Collective Dephasing Quantum Channel). Dephasing is a phenomenon in which the relative phase of a qubit is lost. Quantum channels with collective dephasing act on the input state in the following way:

$$|0\rangle \rightarrow |0\rangle,$$
$$|1\rangle \rightarrow e^{i\phi} |1\rangle,$$

where ϕ is the collective dephasing parameter that varies with time. A logic qubit composed by two physical qubits with anti-parallel parity is immune to collective dephasing, i.e.,

$$|0_L\rangle = |01\rangle,$$
$$|1_L\rangle = |10\rangle.$$

A qubit can be, thus, encoded as $|\psi_L\rangle = \alpha |0_L\rangle + \beta |1_L\rangle$. As expected, $|\psi_L\rangle$ does not suffer from the collective decoherence due to this channel:

$$\mathcal{E}(|\psi_L\rangle) = \mathcal{E}\left(\alpha |0_L\rangle + \beta |1_L\rangle\right)$$
$$= \alpha e^{i\phi} |01\rangle + \beta e^{i\phi} |10\rangle$$
$$= e^{i\phi}\left(\alpha |01\rangle + \beta |10\rangle\right)$$
$$= e^{i\phi} |\psi_L\rangle$$
$$= |\psi_L\rangle ,$$

since the global phase factor $e^{i\phi}$ acquired during this process has no physical significance [7]. It means that the states $|01\rangle$ and $|10\rangle$ belong to $\tilde{\mathcal{H}}$, a decoherence-free subspace from $\mathcal{H}$ in a quantum channel with collective dephasing.

Some practical results already reported in the literature consider the identification, implementation, and adoption of several DFS in quantum computation and communication [2, 17, 29, 31, 33, 35, 41, 51, 57, 59]. For quantum communications, in particular, DFS are useful for building quantum repeaters. Such devices are used for quantum key distribution, quantum teleportation schemes and also for quantum computer networks [13]. The work of Xue [56] shows the characterization of quantum repeaters with DFS for long distance quantum communications.

6.1.1   Method for Obtaining Decoherence-Free Subspaces and Subsystem

Despite the ability to preserve the fidelity of quantum states, one of the limitations regarding the use of DFS relies on the difficulty to identify them [5]. In order to circumvent this problem, Choi and Kribs [9] proposed a method to identify DFS when the error model of the quantum channel is known. The main goal of this section is the characterization of this method that is mainly algebraic.

Let $\mathcal{E} : \mathcal{B}(\mathcal{H}) \to \mathcal{B}(\mathcal{H})$ be a quantum operation. The error model can be specified, for example by the operation elements $\{E_a\}$ of an OSR, $\mathcal{E} \equiv \{E_a\}$. The *noise commutator* $\mathcal{A}'$ for $\mathcal{E}$ is the set of all operators $\mathcal{B}(\mathcal{H})$ which commute with the operators E_a and $E_a^\dagger$. When considering unital channels (which satisfy $\mathcal{E}(\mathbb{1}) = \mathbb{1}$), we have that all $\sigma \in \mathcal{A}'$ satisfy $\mathcal{E}(\sigma) = \sigma$. As a consequence, $\mathcal{A}$ is a †-algebra[1] generated by E_a that is called *interaction algebra* associated with $\mathcal{E}$.

---

[1]The formalism of †-algebras, also known as C^*-algebras, was developed for its use on quantum mechanics of observables. A †-algebra is a Banach $*$-algebra with an additional condition for the norm: $||A^* \cdot A|| = ||A^2||$ for all $A \in \mathcal{U}$, where $\mathcal{U}$ is an algebra with complex norm. A complete tutorial on †-algebras can be found on Davidson [10].

However, quantum channels are generally non-unital and hence we must explore a more general formalism. Any operator σ that belongs to the noise commutator $\mathcal{A}'$ satisfies $\mathcal{E}(\sigma) = \sigma\mathcal{E}(\mathbb{1}) = \mathcal{E}(\mathbb{1})\sigma$. Given a projector P in $\mathcal{B}(\mathcal{H})$, the objective is to find a subalgebra $P\mathcal{B}(\mathcal{H})P$ of $\mathcal{B}(\mathcal{H})$ with algebra $\mathcal{B}(P\mathcal{H})$. To do so, we have the following theorem.

Theorem 6.2 (Choi and Kribs [9]). *Let $\mathcal{E} = \{E_a\}$ be a quantum operation on $\mathcal{B}(\mathcal{H})$. Suppose that P is a projection onto $\mathcal{H}$ that satisfies*

$$\mathcal{E}(P) = P\mathcal{E}(P)P. \qquad (6.6)$$

Then, $E_a P = P E_a P$, $\forall a$. Define

$$\mathcal{A}'_P \equiv \left\{\sigma \in \mathcal{B}(P\mathcal{H}) : [\sigma, PE_a P] = 0 = \left[\sigma, PE_a^\dagger P\right]\right\} \qquad (6.7)$$

and

$$\text{Fix}_P(\mathcal{E}) \equiv \{\sigma \in \mathcal{B}(P\mathcal{H}) : \mathcal{E}(\sigma) = \sigma\mathcal{E}(P) = \mathcal{E}(P)\sigma,$$
$$\mathcal{E}(\sigma^\dagger\sigma) = \sigma^\dagger\mathcal{E}(P)\sigma, \mathcal{E}(\sigma,\sigma^\dagger) = \sigma\mathcal{E}(P)\sigma^\dagger\}. \qquad (6.8)$$

Therefore, $\text{Fix}_P(\mathcal{E})$ is a $\dagger$-algebra inside $\mathcal{B}(P\mathcal{H})$ that coincides with $\mathcal{A}'_P$, i.e.,

$$\text{Fix}_P(\mathcal{E}) = \mathcal{A}'_P. \qquad (6.9)$$

The proof of this theorem will not be fully discussed; we just highlight some of the most important aspects. If P satisfies (6.6), then

$$0 \leq P^\perp E_a P E_a^\dagger P^\perp \leq P^\perp \mathcal{E}(P) P^\perp = 0 \qquad \forall a. \qquad (6.10)$$

To whatever operators $A, B \in \mathcal{B}(\mathcal{H})$, $A \leq B \Rightarrow \langle\psi| B - A |\psi\rangle \geq 0, \forall |\psi\rangle \in \mathcal{H}$. This way, $P^\perp E_a P = 0$ or, equivalently, $E_a P = P E_a P$, $\forall a$. When considering $\sigma \in \mathcal{A}'_P$, then

$$\mathcal{E}(\sigma) = \sum_a E_a P \sigma P E_a^\dagger$$

$$= \sigma \sum_a E_a P E_a^\dagger = \sum_a E_a P E_a^\dagger \sigma$$

$$= \sigma\mathcal{E}(P) = \mathcal{E}(P)\sigma. \qquad (6.11)$$

Projectors P satisfying (6.6) have some properties. For instance, a quantum channel $\mathcal{E} \equiv \{E_a\}$ acts on a quantum state $\sigma \in \mathcal{A}'_P$ projecting it into another state σ' in the subspace defined by P. To support this statement, we have that

$$\sigma' = \mathcal{E}(\sigma)$$
$$= \sigma\mathcal{E}(P)$$
$$= (P\sigma P)(P\mathcal{E}(P)P)$$
$$= P[\sigma P\mathcal{E}(P)]P \in \mathcal{B}(P\mathcal{H}). \tag{6.12}$$

In this particular case, $\mathcal{E}(\sigma) = \sigma$ only if $\mathcal{E}(P) = \mathbb{1}$.

The next step is to show how projectors with such characterization can capture the DFS of a quantum operation $\mathcal{E}$.

Theorem 6.3 (Method for Obtaining DFS [9]). *Let $\mathcal{E}$ be a quantum operation in $\mathcal{B}(\mathcal{H})$. Let P be a projector that satisfies (6.6), and let $P\mathcal{H} = \oplus_k(\mathcal{H}^{A_k} \otimes \mathcal{H}^{B_k})$ be the decomposition of $P\mathcal{H}$ induced by the structure of the †-algebra $\mathcal{A}'_P = \mathrm{Fix}_P(\mathcal{E})$. Then, the subsystems $\mathcal{H}^{B_k}$, with $\dim(\mathcal{H}^{B_k}) > 1$, are decoherence-free for $\mathcal{E}$.*

We can say that the essence of this method relies on the identification of all projectors P satisfying (6.6). Thenceforth, the structure of $\mathcal{A}'_P = \mathrm{Fix}_P(\mathcal{E})$ is used to determine what are the states that belong to the DFS.

One important aspect is the *optimality* of the proposed method. It means that it can capture all projectors satisfying (6.6) [9, Theorem 3]. Despite the characterization of such method, the authors state that no computational procedures were developed to this purpose yet.

Example 6.2 (Identifying a DFS in a Quantum Channel). Suppose that the quantum channel $\mathcal{E} \equiv \{E_0, E_1, E_2\}$ acts on a bidimensional space state with the following Kraus operators:

$$E_0 = \alpha(|00\rangle\langle00| + |11\rangle\langle11|) + |01\rangle\langle01| + |10\rangle\langle10|,$$
$$E_1 = \beta(|00\rangle\langle00| + |11\rangle\langle11| + |01\rangle\langle01| + |10\rangle\langle10|),$$
$$E_2 = \beta(|00\rangle\langle00| + |11\rangle\langle11| - |01\rangle\langle01| - |10\rangle\langle10|),$$

where q is a scalar, $0 < q < 1$; $\alpha = \sqrt{1 - 2q}$; $\beta = \sqrt{q/2}$. It is possible to notice that $\mathcal{E}(\mathbb{1}) = \sum_{a=0}^{2} E_a E_a^\dagger \neq \mathbb{1}$ and, therefore, this channel is not unital.

In this channel model, there is only one state ρ such that $\mathcal{E}(\rho) = \rho$. However, such invariance does not come from the action of $\mathcal{E}$, but from a fixed point. Despite that, there is a DFS with such dimension when we consider the projector $P = |01\rangle\langle01| + |10\rangle\langle10|$, i.e., all operators supported by P are invariant under $\mathcal{E}$. It means that $\mathcal{E}(\sigma') = \sigma'$ for all $\sigma' = P\sigma P$.

To exemplify this statement, let the density operator of the state $|\psi\rangle$ be

$$|\psi\rangle\langle\psi| = \frac{|01\rangle\langle01| + |01\rangle\langle00| + |00\rangle\langle01| + |00\rangle\langle00|}{2}.$$

Applying the projector P onto $|\psi\rangle$ results

$$|\psi'\rangle\langle\psi'| = P|\psi\rangle\langle\psi|P$$

$$= (|01\rangle\langle01| + |10\rangle\langle10|)\left(\frac{|01\rangle\langle01| + |01\rangle\langle00| + |00\rangle\langle01| + |00\rangle\langle00|}{2}\right)P$$

$$= \left(\frac{|01\rangle\langle01| + |01\rangle\langle00|}{2}\right)(|01\rangle\langle01| + |10\rangle\langle10|)$$

$$= \frac{|01\rangle\langle01|}{2}.$$

Note that $|\psi'\rangle\langle\psi'|$ does not vary after passing the channel $\mathcal{E}$, despite it does not belong to the noise commutator $\mathcal{A}'$ for $\mathcal{E}$:

$$\mathcal{E}(|\psi'\rangle\langle\psi'|) = \sum_{a=0}^{2} E_a |\psi'\rangle\langle\psi'| E_a^\dagger$$

$$= \frac{|01\rangle\langle01|}{2} + \beta \cdot \frac{|01\rangle\langle01|}{2} - \beta \cdot \frac{|01\rangle\langle01|}{2}$$

$$= \frac{|01\rangle\langle01|}{2}.$$

6.1.2 Relation with the Zero-Error Capacity of Quantum Channels

The work of Medeiros et al. [39] explores the relation between DFS and zero-error capacity of quantum channels. This relation is established from the method for obtaining DFS of Choi and Kribs [9], showed in the previous section. The purpose of this section is to show this relation.

We know that a quantum channel has zero-error capacity if and only if there are at least two non-adjacent states at the channel input. Considering an optimum pair $(\mathcal{S}, \mathcal{M})$ according to Definition 5.3, it is possible to derive a pair $(\mathcal{S}', \mathcal{M}')$, where $\mathcal{S}' \subset \mathcal{S}$, $\mathcal{M}' = \{M_1, \ldots, M_k, M_{k+1}\} \subset \mathcal{M}$, and $M_{k+1} = \mathbb{1} - \sum_{i=1}^{k} M_i$. The projectors $M_i \in \mathcal{M}'$, with $1 \le i \le k$, satisfy

$$\mathcal{E}(M_i) = M_i \mathcal{E}(M_i) M_i \tag{6.13}$$

and

$$M_i M_j = \delta_{ij} M_i M_j, \tag{6.14}$$

where δ denotes the Kronecker's delta. When choosing projectors with such restrictions, we notice that the elements of $\mathcal{S}'$ can define a DFS, as established in the method for obtaining DFS explored in Sect. 6.1.1.

As a consequence, we have that the set $(\mathcal{S}', \mathcal{M}')$ is optimum and the zero-error capacity $C^{(')}(\mathcal{E})$ defined for this set can be bigger than the zero-error capacity $C^{(0)}(\mathcal{E})$ defined for $(\mathcal{S}, \mathcal{P})$, it means that $C^{(')}(\mathcal{E}) \geq C^{(0)}(\mathcal{E})$. The proofs of such consequences make use of graph theory and mappings properties [39].

In summary, the conclusion of those authors regarding the relation of DFS and zero-error capacity is that if a zero-error quantum channel has a DFS, then the zero-error capacity must be obtained from the DFS by using projectors that attain certain properties.

6.2 Quantum Secrecy Capacity

The privacy in quantum systems was initially considered by Schumacher and Westmoreland [45]. These researchers conceived a model that allows two legitimate parties, Alice and Bob, to exchange classical messages through a noisy quantum channel. An eavesdropper (Eve) has total access to the environment of the quantum channel from which she is able to capture information of the legitimate parties.

Alice sends messages from a set of integers $\mathcal{U} = \{1, 2, \dots, |\mathcal{U}|\}$ mapped on an ensemble of quantum states $\{\rho(u), p_u : u \in \mathcal{U}\}$. The states of the ensemble are called *quantum codewords*, composed by tensor products of quantum states:

$$\rho(u) = \rho_1(u) \otimes \rho_2(u) \otimes \dots \otimes \rho_n(u) \quad u \in \mathcal{U}, \rho_i(u) \in \mathcal{H}, i = 1, 2, \dots, n. \quad (6.15)$$

The mapping characterizes a quantum block code with block length n and rate $R = \frac{1}{n} \log |\mathcal{U}|$. A decoding scheme for this quantum code is a decoding function that associates univocally an output quantum state with a set of integers, i.e., $g : \mathcal{H} \to \mathcal{U}$, $\hat{u} = g(\mathcal{E}(\rho(u))) \in \mathcal{U}$. An error occurs when $g(\mathcal{E}(\rho(u))) \neq u$.

The quantum privacy between Alice and Bob is limited by the *coherent information* among them. The coherent information is an information measure that quantifies the difference between the von Neumann entropies of two systems: the system of interest and the environment [45]. When considering this formulation, Cai et al. [6] and Devetak [11] notice some similarities with classical wiretap channels proposed by Wyner [54]. Then, they proposed a quantum version of such channels, presented in Definition 6.3 and illustrated in Fig. 6.1.

Definition 6.3 (Quantum Wiretap Channel). A quantum memoryless wiretap channel is described by a superoperator $\mathcal{E}$ in a complex Hilbert space $\mathcal{H} = \mathcal{H}_{\text{Bob}} \otimes \mathcal{H}_{\text{Eve}}$. When Alice sends a quantum state $\rho \in \mathcal{H}^{\otimes n}$, Bob receives $\rho_{\text{Bob}} = \text{Tr}_{\text{Eve}}[\mathcal{E}^{\otimes n}(\rho)]$ and Eve receives $\rho_{\text{Eve}} = \text{Tr}_{\text{Bob}}[\mathcal{E}^{\otimes n}(\rho)]$, where n is the dimension of input Hilbert space.

When communicating over a quantum wiretap channel, security can be achieved by using a particular type of quantum block code: the *quantum wiretap codes*. Two additional parameters are necessary: λ, which represents an upper bound for the error-probability; and μ, which represents an upper bound for the maximum

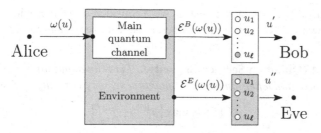

Fig. 6.1 General model of a quantum wiretap channel

accessible information by the eavesdropper Eve. A quantum wiretap code is referred to as a 4-tuple $(n, |\mathcal{U}|, \lambda, \mu)$. A formal characterization of such codes is given below.

Definition 6.4 (Quantum Wiretap Block Codes). Consider a quantum block code of length n and rate $R = \frac{1}{n} \log |\mathcal{U}|$, where $\mathcal{U} = \{1, 2, \ldots, |\mathcal{U}|\}$ is a set of classical messages. The set of codewords labeled by the index of the messages is given as follows:

$$\Omega(\mathcal{U}) = \{\rho(u) : u \in \mathcal{U}\}. \tag{6.16}$$

We assume that the decoding function is given by the POVM $\{\mathcal{D}_u : u \in \mathcal{U}\}$, where $\sum_u \mathcal{D}_u \leq \mathbb{1}$.

This code is said to be a quantum wiretap block code with parameters $(n, |\mathcal{U}|, \lambda, \mu)$, or quantum wiretap code for short, if two conditions are attained:

$$P_e = 1 - \frac{1}{|\mathcal{U}|} \sum_{u \in \mathcal{U}} \mathrm{Tr}_{\mathrm{Eve}}[\mathcal{E}(\rho(u))\mathcal{D}_u] \leq \lambda, \tag{6.17}$$

and

$$\frac{1}{n} \left\{ S \left(\sum_{u \in \mathcal{U}} \mathrm{Tr}_{\mathrm{Bob}}[\mathcal{E}^{\otimes n}(\rho(u))] \right) - \sum_{u \in \mathcal{U}} \frac{1}{|\mathcal{U}|} S(\mathrm{Tr}_{\mathrm{Bob}}[\mathcal{E}^{\otimes n}(\rho(u))]) \right\} \leq \mu. \tag{6.18}$$

In the definition of a quantum wiretap code with parameters $(n, |\mathcal{U}|, \lambda, \mu)$, (6.17) ensures an average probability of decoding errors for Bob lower than λ, and (6.18) limits the information accessible to the eavesdropper, which captures almost nothing from the message sent by Alice [6].

Lastly, the *secrecy capacity of a quantum channel* is defined as follows.

Definition 6.5 (Quantum Secrecy Capacity). The secrecy capacity of a quantum channel $\mathcal{E}$ is the largest real number $C_S(\mathcal{E})$, such that for all $\epsilon, \lambda, \mu > 0$ and n large enough, there is a quantum wiretap code with parameters $(n, |\mathcal{U}|, \lambda, \mu)$ such that

$$C_S(\mathcal{E}) < \frac{1}{n} \log |\mathcal{U}| + \epsilon. \tag{6.19}$$

Only uniformly distributed messages were considered in the previous definitions, but the following theorem is a more general result for the quantum secrecy capacity [6, Sect. 5].

Theorem 6.4 (Quantum Secrecy Capacity). *Let $\mathcal{E}$ be a quantum wiretap channel as characterized in Definition 6.3. The quantum secrecy capacity of $\mathcal{E}$ satisfies*

$$C_S(\mathcal{E}) \geq \max_{\{P\}} \left[\chi^{Bob} - \chi^{Eve} \right], \tag{6.20}$$

where the maximum is taken over all probability distributions over $\mathcal{U}$; and χ^{Bob} and χ^{Eve} are Holevo quantities defined as

$$\chi^{Bob} = S(\rho_{Bob}) - \sum_i p_i S(\rho_{Bob}(i)), \tag{6.21}$$

$$\chi^{Eve} = S(\rho_{Eve}) - \sum_i p_i S(\rho_{Eve}(i)), \tag{6.22}$$

where ρ_{Bob} is the state received by Bob after a partial trace over the environment; and ρ_{Eve} is Eve's final state.

The proof of this theorem makes use of the *random coding proof* technique to ensure that the information gathered by Eve is negligible. When the information transmission rate through the channel is smaller than the quantum secrecy capacity, the protocol guarantees *unconditional security* [6]. This capacity is equivalent to the definition of privacy presented by Schumacher and Westmoreland [45].

The quantum secrecy capacity (6.20) is the quantum counterpart of the classical secrecy capacity proposed by Wyner [54]. We can, therefore, notice some similarities between both definitions: they limit the decoding error probability and the information accessible to the wiretapper.

A particular characteristic of the quantum secrecy capacity is that it does not have single letter characterization, i.e., the capacity cannot be directly calculated because the maximum is taken over all possible input states as well as all possible probability distributions [6, 11].

Some codes for quantum wiretap channels can be found in the literature. Hamada [25, 26] proposed classes of codes for both classical and quantum wiretap channels. In the quantum case, they are based on concatenated conjugate codes that are equivalent to the Calderbank-Shor-Steane (CSS) codes [42, Sect. 10.4.2]. Another characteristic of the proposed code is the polynomial-time complexity for encoding and decoding in terms of channel usage.

Another class of codes for quantum wiretap channels was proposed by Wilde and Guha [53]. This construction is based on polar codes for degraded wiretap channels that reach the symmetric secrecy capacity for a quantum wiretap channel with a classical eavesdropper. Although this class of codes also has a polynomial-time complexity for encoding and decoding, examples of such codes are strongly dependent on numerical simulations [16]. Nonetheless, the authors showed that such codes perform well when used to carry information through amplitude damping, dephasing, erasure, and cloning quantum channels [16, 53].

6.3   Zero-Error Secrecy Capacity

Consider a scenario where two legitimate parties, Alice and Bob, want to exchange classical messages through a quantum channel $\mathcal{E}$ in a secret and error-free way. These messages must be protected from an eavesdropper (Eve), which has complete and non-restricted access to the environment. This communication model is similar to the scenario already considered in Fig. 6.1.

The communication model where the eavesdropper has complete access to the environment follows the formalism proposed by Cai et al. [6] and of Devetak [11] for the characterization of quantum wiretap channels. In practical scenarios, it is more common to consider a direct action of the eavesdropper on the main quantum channel and its implications in the communication and in the non-authorized information gathering, e.g., in quantum key distribution protocols. The scenario described in Fig. 6.1, although different from this approach, can also be physically implemented and is already consolidated in the literature for quantum privacy purposes [45]. In particular, the channel $\mathcal{E}$ has Kraus operators $\{E_a\}$ and positive zero-error capacity. The following characterization presents the quantum channel under consideration.

Characterization 6.1 (Quantum Channel with Positive Zero-Error Capacity). *Let $\mathcal{E}$ be a trace-preserving quantum map with Kraus operators $\{E_a\}$, which represents a noisy quantum channel $\mathcal{E}$. We consider that $\mathcal{E}$ has a strictly positive zero-error capacity, $C^{(0)}(\mathcal{E}) > 0$, reached by an optimum pair $(\mathcal{S}, \mathcal{M})$.*

If there exists a POVM $\mathcal{M}' = \{M_1, \ldots, M_k\}$ that satisfies (6.13) and (6.14), then

$$\mathcal{E}(M_i) = M_i \mathcal{E}(M_i) M_i, \tag{6.23}$$

$$M_i M_j = \delta_{i,j} M_i M_j, \tag{6.24}$$

for all $i, j \leq k$. Furthermore, if we define

$$\mathcal{S}' = \left\{ \rho_i = |s_i\rangle \langle s_i|_{i=1}^k, \rho_i \in M_i \mathcal{H} \text{ and } [\rho_i, M_i E_a M_i] = 0 = [\rho_i, M_i E_a^\dagger M_i] \right\}, \tag{6.25}$$

then the pair $(\mathcal{S}', \mathcal{M}')$ is also optimum. Since $(\mathcal{S}', \mathcal{M}')$ has been obtained according to the method described in Sect. 6.1.1, the quantum states $\rho_i \in \mathcal{S}'$ characterize an orthonormal basis set for the decoherence-free subspace $\tilde{\mathcal{H}}$. For the sake of simplicity, from now on we will use the notation $\tilde{\mathcal{H}}$ in a reference for the basis states of this decoherence-free subspace. Therefore, states in $\mathcal{S}'$ can be used to encode information that will be immune to an eavesdropper, as shown in the following lemmas.

Lemma 6.1 (Optimum Pair $(\mathcal{S}', \mathcal{M}')$ Defines a QEAC). *The optimum pair $(\mathcal{S}', \mathcal{M}')$ is a quantum error avoiding code (vide Sect. 6.2).*

Proof. In order to prove this lemma, we must show that the pair $(S', \mathcal{M}')$ has all the elements of a QEAC.

Let $\mathcal{U} = \{u_1, \ldots, u_k\}$ be a set of classical messages; each message in $\mathcal{U}$ is associated with a state in S' through a bijection. The set S' defines a codebook $\tilde{P}(\mathcal{U}) = \{\tilde{\rho}(u_i) = \rho_i\} \equiv S'$ with codewords of length n. The decoding is performed by a set of positive operators $M_i \in \mathcal{M}'$, $i \in 1, \ldots, |\mathcal{U}|$, with $\sum_{i=1}^{|\mathcal{U}|} M_i \leq \mathbb{1}$. Indeed, there is a bijective correspondence between the set of POVM operators $\mathcal{M}'_i$ and the set of messages $\mathcal{U}$. Therefore, the pair $(\tilde{P}(\mathcal{U}), \mathcal{M}')$, which is equivalent to $(S', \mathcal{M}')$, defines a quantum error-avoiding code of length n and rate $\frac{1}{n} \log |\mathcal{U}|$.

It is straightforward to see that for each DFS we can construct a quantum error-avoiding code to the corresponding quantum channel. The channel $\mathcal{E}$ is subject to collective decoherence, being governed by the Hamiltonian (6.1). Thanks to symmetries existing in collective decoherence, states in the DFS do not suffer the action of $\mathbb{H}_{SE}$, the Hamiltonian component representing the interaction between system and environment.

When Alice wants to send a message u to Bob using a quantum error-avoiding code, she encodes u into a quantum codeword $\tilde{\rho}(u)$ and sends the corresponding state through the channel $\mathcal{E}$. We assume that the environment starts in a pure state $|0_E\rangle \langle 0_E|$. Due to the decoherence, Bob and Eve will receive the following states, respectively,

$$\rho_{\text{Bob}}(\tilde{\rho}(u)) = \text{Tr}_{\text{Eve}} \left[\mathcal{E}(\tilde{\rho}(u) \otimes |0_E\rangle \langle 0_E|) \right], \qquad (6.26)$$

$$\rho_{\text{Eve}}(\tilde{\rho}(u)) = \text{Tr}_{\text{Bob}} \left[\mathcal{E}(\tilde{\rho}(u) \otimes |0_E\rangle \langle 0_E|) \right]. \qquad (6.27)$$

Since Alice uses a QEAC, dynamic symmetries protect the quantum codeword from interacting with the environment. Therefore, the joint evolution between system and environment happens in a decoupled way. Thus, the state $\rho_{\text{Bob}}(\tilde{\rho}(u))$ is given by

$$\rho_{\text{Bob}}(\tilde{\rho}(u)) = \text{Tr}_{\text{Eve}} \left[\mathcal{E}(\tilde{\rho}(u) \otimes |0_E\rangle \langle 0_E|) \right] \qquad (6.28)$$

$$= \text{Tr}_{\text{Eve}} \left[\sum_a E_a (\tilde{\rho}(u) \otimes |0_E\rangle \langle 0_E|) E_a^\dagger \right]$$

$$= \text{Tr}_{\text{Eve}} \left[\tilde{\rho}(u) \otimes \rho_E \right] \qquad (6.29)$$

$$= \tilde{\rho}(u), \qquad (6.30)$$

where (6.29) is due to the invariance of a state from a DFS under the OSR operators.

Taking into account the Hamiltonian (6.1) of the quantum system and considering the fact that system of interest and environment have not interacted, then it is possible to ensure that the environment suffered only the action of $\mathbb{H}_E$, which indicates a unitary evolution restricted to the environment. It means that $\rho_{\text{Eve}}(\tilde{\rho}(u)) = \rho_E$ (6.27) is a *pure state*.

Proceeding with the development, it is possible to state and prove the following lemma.

Lemma 6.2 (Optimum Pair $(S', \mathcal{M}')$ Defines a Wiretap Code). *The pair $(S', \mathcal{M}')$ defines a quantum wiretap code with parameters $(n, |\mathcal{U}|, 0, 0)$.*

Proof. In Definition 6.4 of a wiretap code as proposed by Cai et al. [6], two conditions must be satisfied in order to achieve secrecy: (1) the average error decoding probability must be small; and (2) the accessible information to the eavesdropper must be arbitrarily small. As we show below, these two requirements are actually satisfied.

For the first condition, note that the pair $(S', \mathcal{M}')$ is optimal, i.e., the set S' attains the zero-error capacity. If quantum codewords are composed of tensor products of states in S', then the communication is accomplished without decoding errors. Therefore, $\lambda = 0$ and the first condition is attained.

In order to verify the second condition, we need to check the accessible information by Eve, which is given as

$$S\left(\sum_{u \in \mathcal{U}} \frac{1}{|\mathcal{U}|} \mathrm{Tr}_{\mathrm{Bob}} \, \mathcal{E}(\tilde{\rho}(u))\right) - \sum_{u \in \mathcal{U}} \frac{1}{|\mathcal{U}|} S\left(\mathrm{Tr}_{\mathrm{Bob}} \, \mathcal{E}(\tilde{\rho}(u))\right) \le \mu, \tag{6.31}$$

where μ is arbitrarily small. Instead of calculating the left side of (6.31), we make use of an upper bound for the accessible information, the Holevo quantity, defined by

$$\chi^{\mathrm{Eve}} = S(\rho_{\mathrm{Eve}}(\tilde{\rho}(u))) - \sum_u p_u S(\rho_{\mathrm{Eve},u} \tilde{\rho}(u)). \tag{6.32}$$

Because quantum codewords are composed by states in S' that belongs to a DFS, there are no interactions between the system and the environment. Therefore, the initial environment state, $|0_E\rangle$, evolves only under the Hamiltonian $\mathbb{H}_E$, indicating a unitary evolution restricted to the environment. It means that the final environment state is pure. This way:

$$\chi^{\mathrm{Eve}} = S(\rho_{\mathrm{Eve}}(\tilde{\rho}(u))) - \sum_u p_u S(\rho_{\mathrm{Eve},u} \tilde{\rho}(u))$$

$$= S(\rho_E) - \sum_u p_u S(\rho_{\mathrm{Eve},u} \tilde{\rho}(u))$$

$$= 0 - \sum_u p_u S(\rho_{\mathrm{Eve},u} \tilde{\rho}(u)). \tag{6.33}$$

Because $\chi^{\mathrm{Eve}} \ge 0$, $S(\rho) \ge 0$ for any ρ, and that $p_u \ge 0$ for all u, then the sum at the right side of (6.33) is zero. Therefore, $\chi^{\mathrm{Eve}} = 0$. Since the Holevo quantity is an upper bound for accessible information, the left side of (6.31) is zero, i.e., $\mu = 0$.

Lemmas 6.1 and 6.2 guarantee that unconditionally secure communication can be performed by using codewords composed by quantum states belonging to a DFS for corresponding quantum channel [23].

Even though an optimum (S', M') defines a wiretap code with parameters $(n, |U|, 0, 0)$, it is not always possible to extract (S', M') from an optimum pair (S, M). According to Lemma 6.1, the quantum states in S' belongs to $\tilde{H}$ for the channel $\mathcal{E}$.

However, considering practical scenarios, a DFS may exist in such conditions, even with smaller cardinality than the set of messages, i.e., with $dim(\tilde{H}) < |U|$. For such situations, we use a wiretap code with parameters $(n, dim(\tilde{H}), 0, 0)$, which allows a communication free of errors and without information leakage. Despite that, in this second situation the communication occurs with a lower rate than when considered the code obtained according to the conditions previously mentioned. Taking this into account and also both lemmas proved, we can characterize a new kind of capacity for quantum channels, whose definition is given as follows.

Definition 6.6 (Zero-Error Secrecy Capacity). Let $\mathcal{E}$ be a quantum channel according to Characterization 6.1. We define the zero-error secrecy capacity of $\mathcal{E}$ as the largest real number $C_S^{(0)}(\mathcal{E})$ such that, for every $\epsilon > 0$ and sufficiently large n, there is a quantum wiretap code $(n, |U|, 0, 0)$ which satisfies

$$C_S^{(0)}(\mathcal{E}) \leq \frac{1}{n} \log |U| + \epsilon. \tag{6.34}$$

Two main features of this capacity are the absence of decoding errors and of information leakage to the eavesdropper. It is in contrast with the secrecy capacity of quantum channels, in which decoding errors among the legitimate parties can occur.

The following theorem gives a way of quantifying the zero-error secrecy capacity.

Theorem 6.5 (Zero-Error Secrecy Capacity). *Let $\mathcal{E}$ be a quantum channel according to Characterization 6.1. The zero-error secrecy capacity of $\mathcal{E}$ is given by*

$$C_S^{(0)}(\mathcal{E}) \equiv \min \left\{ C^{(0)}(\mathcal{E}), C_S(\mathcal{E}) \right\} \tag{6.35}$$

$$\equiv \min \left\{ \sup_{\tilde{H}} \sup_{n} \frac{1}{n} \log \dim(\tilde{H})^n, \max_{\{P\}} \chi^{Bob} \right\}, \tag{6.36}$$

where n is the length of the code; the maximum is taken over all probability distributions P over U, and χ^{Bob} denotes an upper bound for the accessible information of the receiver (Bob):

$$\chi^{Bob} = S\left(\sum_{u} p_u \rho_{Bob}(\tilde{\rho}(u)) \right) - \sum_{u} p_u S\left(\rho_{Bob}(\tilde{\rho}(u)) \right), \tag{6.37}$$

where p_u is the a priori probability of the symbol $u \in U$.

Proof. This proof considers some facts about the capacities of a quantum channel $\mathcal{E}$. Let $C_{1,\infty}(\mathcal{E})$ be the ordinary classical capacity of $\mathcal{E}$ defined according to the Holevo-Schumacher-Westmoreland theorem [28, 44]. Let $C_S(\mathcal{E})$ be the secrecy capacity of $\mathcal{E}$ [6, 11]. And, lastly, let $C^{(0)}(\mathcal{E})$ be the classical zero-error capacity of a quantum channel $\mathcal{E}$ [38]. We have that $C_S(\mathcal{E}) \leq C_{1,\infty}(\mathcal{E})$, and that $C^{(0)}(\mathcal{E}) \leq C_{1,\infty}(\mathcal{E})$.

Considering that $|\mathcal{U}| = \dim(\tilde{\mathcal{H}})$, a code with parameters $(n, |\mathcal{U}|, 0, 0)$ is simultaneously an error-free code and also a wiretap code. By definition, we know that the zero-error capacity is related to the maximum amount of messages that are distinguishable at the channel output. Since each word in the alphabet was associated with a state of a DFS, according to Lemma 6.1, we have

$$C^{(0)}(\mathcal{E}) = \sup_{\tilde{\mathcal{H}}} \sup_n \frac{1}{n} \log \dim(\tilde{\mathcal{H}})^n, \qquad (6.38)$$

where n is the length of the code. Since this is a wiretap code having input symbols belonging to $\tilde{\mathcal{H}}$, $C_S(\mathcal{E}) = \chi^{\text{Bob}} - \chi^{\text{Eva}}$. As a consequence of Lemma 6.2,

$$C_S^{(0)}(\mathcal{E}) \geq \max_{\{P\}} \left[\chi^{\text{Bob}} - \chi^{\text{Eva}} \right]$$

$$\geq \max_{\{P\}} \left[\chi^{\text{Bob}} - 0 \right]$$

$$= \max_{\{P\}} \chi^{\text{Bob}}, \qquad (6.39)$$

where the maximum is taken over all a priori probability distributions P of the symbols $u \in \mathcal{U}$. The equality follows from the HSW theorem. We have to consider two situations:

1. There exists an optimum pair $(\mathcal{S}', \mathcal{M}')$ derived from $(\mathcal{S}, \mathcal{M})$ according to (6.23) and (6.24). In this case, $|\mathcal{U}| = \dim(\tilde{\mathcal{H}})$ and $C_S^{(0)}(\mathcal{E}) = C_S(\mathcal{E}) = C^{(0)}(\mathcal{E})$.
2. There exists a DFS $\tilde{\mathcal{H}}$ for the channel that is not directly obtained from the error-free code. In this situation, $C_S(\mathcal{E}) < C^{(0)}(\mathcal{E})$, i.e., error-free and leakage-free communication is only possible if $C_S^{(0)}(\mathcal{E}) = \min\{C^{(0)}(\mathcal{E}), C_S(\mathcal{E})\}$.

This way, the final expression for the zero-error secrecy capacity can be described in terms of the relation between the zero-error capacity and the secrecy capacity:

$$C_S^{(0)}(\mathcal{E}) = \min\{C^{(0)}(\mathcal{E}), C_S(\mathcal{E})\}, \qquad (6.40)$$

where $C^{(0)}(\mathcal{E})$ and $C_S(\mathcal{E})$ are the zero-error capacity and the secrecy capacity of $\mathcal{E}$, respectively.

When a quantum channel $\mathcal{E}$ has $C_S^{(0)}(\mathcal{E}) = \sup_{\tilde{\mathcal{H}}} \sup_n \frac{1}{n} \log \dim(\tilde{\mathcal{H}})^n$, then the zero-error secrecy capacity is straightforwardly obtained from the dimension of the largest existing DFS for the channel.

According to Medeiros et al. [40], the zero-error capacity can be achieved using tensor product of pure states at the channel input. We can see that the same holds for the zero-error secrecy capacity $C_S^{(0)}(\mathcal{E})$.

The zero-error secrecy capacity communication protocol has the same level of security of the protocol established by Schumacher and Westmoreland [45]. According to the authors, the ability of a quantum channel to send private information is at least as great as its ability to send coherent information. In the zero-error secrecy capacity scenario, the information can be retrieved completely free of errors at the channel output. Therefore, the ability to communicate private information is maximized.

When considering the difficulties to implement quantum channels that enable communications completely free of errors [34], the zero-error secrecy capacity allows error-free and secure communications to be performed since the quantum channel attains some conditions. This is the case of quantum channels with collective decoherence [13, 30, 55], and the quantum channels with positive zero-error capacity discussed in [24]. In the latter example, the quantum channel proposed by Xue [56] can be used for long-distance zero-error quantum communications.

Although the zero-error secrecy capacity was adequately defined, it is zero for many kinds of quantum channels. We can say, indeed, that this capacity is different from zero only for quantum channels with positive zero-error capacity and for channels under the effect of collective-decoherence, allowing the existence of decoherence-free subspaces. Nevertheless, the definition of the zero-error secrecy capacity can improve our knowledge regarding the "abilities" of quantum channels, allowing a more adequate use in certain situations.

6.4   Representation in Graphs

In this section the relation between the ZESC and the graph theory will be depicted. Unfortunately, this relation is not so general as for the zero-error capacity of quantum channels, as presented previously in Sect. 5.2. The relation is only useful to describe quantum channels satisfying the first situation described in the proof of Theorem 6.5.

If there is a non-empty subset $\mathcal{M}'$ obtained from $\mathcal{M}$ according to (6.23) and (6.24), then it follows from the method of Choi and Kribs shown previously in Sect. 6.1.1 that $(\mathcal{S}', \mathcal{M}')$ characterizes a DFS $\tilde{\mathcal{H}}$, which is a subspace of the input Hilbert space $\mathcal{H}$. Supposing the existence of a set $\mathcal{S}'$, it is possible to build a characteristic graph for quantum channels with positive zero-error secrecy capacity. This construction is similar to that made for the zero-error capacity, as presented in Definition 5.5. However, there are some differences between the two vertex sets in each case.

Let $\mathcal{E}$ be a quantum channel with positive zero-error secrecy capacity attaining the first situation of the Theorem 6.5. The characteristic graph of $\mathcal{E}$, denoted by $\tilde{\mathcal{G}} = \langle V, E \rangle$, is built as follows.

1. The vertex set V is composed by the elements $\tilde{\mathcal{H}}$, which are referred by the indexes of the corresponding messages, i.e., $V = \{1, 2, \ldots, \dim(\tilde{\mathcal{H}})\}$.
2. The set of edges E connects two vertices if they are non-adjacent at the channel's end (see Definition 5.4).

The n-th Shannon product of $\tilde{\mathcal{G}}$, denoted by $\tilde{\mathcal{G}}^n$, has the vertex set V^n, each vertex corresponding to an n-tensor product of state belonging to $\mathcal{S}'^{\otimes n}$. Two vertices in V^n are connected if the two corresponding n-tensor product states are adjacent.

Taking under consideration such graph, since the elements of a DFS $\tilde{\mathcal{H}}$ are pairwise distinguishable at the channel's end, then the resulting graph is *complete*. Thus, the largest number of messages that can be transmitted without error by the quantum channel $\mathcal{E}$ is given by the clique number $\tilde{\mathcal{G}}^n$.

This way, the zero-error secrecy capacity of a quantum channel $\mathcal{E}$ that attends the situation 1 of Theorem 6.5 is

$$C_S^{(0)} = \sup_{\tilde{\mathcal{H}}} \sup_n \frac{1}{n} \log \omega(\tilde{\mathcal{G}}^n). \tag{6.41}$$

Given a certain integer and a graph, finding a clique in the graph with size equal to the integer given is an $\mathcal{NP}$-Complete problem. However, some characteristics of the zero-error and of DFS can be taken into account to obtain $C_S^{(0)}(\mathcal{E})$. If the graph built from $\tilde{\mathcal{H}}$ is complete, then the clique number $\tilde{\mathcal{G}}$ is equal to $\dim(\tilde{\mathcal{H}})$, which takes us to the known expression (6.38). Such relation between the clique number and the cardinality of the corresponding set of vertices does not arise in ordinary quantum zero-error channels. This particularity arises thanks to the DFS.

6.5 Security Analysis

To analyze the security of the proposed scheme, we have to consider that there are three types of secrecy.

1. **Strong Secrecy.** It requires that the total amount of information transferred to the eavesdropper goes to zero in the asymptotic limit of the number of communications;
2. **Weak Secrecy.** It requires that the information per symbol transferred to the eavesdropper go to zero in the asymptotic limit of the number of communications [50];
3. **Perfect Secrecy.** It requires that no information is transferred to the eavesdropper [48].

According to the communication scheme proposed, when Alice encodes a message using a quantum wiretap code with parameters $(n, |\mathcal{U}|, 0, 0)$ and sends it to Bob, we have that the set of input states belong to a DFS. Thanks to the DFS, the input states do not interact with the environment. The eavesdropper, in turn, has access only to the environment whose state is pure along the interaction. As a consequence, the information accessible to Eve is zero, obtained from $\chi^{\text{Eva}} = 0$ as shown in the proof of Lemma 6.2. Eve's uncertainty regarding the secret messages does not have changes, even if she observed the state of the environment completely. We can conclude, therefore, that the scheme under consideration has perfect secrecy.

6.6   Examples

We will now show some examples regarding the zero-error secrecy capacity.

Example 6.3 (Strictly Positive ZESC). Initially, we assume that a quantum channel $\mathcal{E}_1$ has positive quantum zero-error capacity reached by an optimum pair $(\mathcal{S}_1, \mathcal{M}_1)$, as shown in Fig. 6.2a. By following the procedures described in Sect. 6.3, a pair $(\mathcal{S}_1', \mathcal{M}_1')$ is obtained, as shown in Fig. 6.2b.

Characteristic graphs for $\mathcal{E}_1$ with inputs $(\mathcal{S}_1, \mathcal{M}_1)$ and $(\mathcal{S}_1', \mathcal{M}_1')$ can be found in Fig. 6.3a, b, respectively.

As can be seen, the largest clique has size 2 and is obtained by the pair $(0, 1)$ in both cases. It leads to a quantum zero-error capacity equal to

$$C^{(0)}(\mathcal{E}_1) = \sup_{\tilde{\mathcal{H}}_1} \sup_n \frac{1}{n} \log \dim(\tilde{\mathcal{H}}_1)^n$$

$$= \log 2$$

$$= 1 \text{ bit per symbol per channel use.} \qquad (6.42)$$

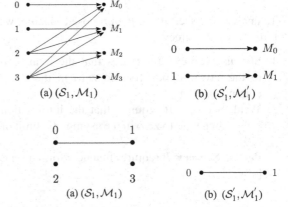

Fig. 6.2 Representation of the transitions performed in the quantum channel $\mathcal{E}_1$ for input states from optimum pairs (**a**) $(\mathcal{S}_1, \mathcal{M}_1)$ and (**b**) $(\mathcal{S}_1', \mathcal{M}_1')$

(a) $(\mathcal{S}_1, \mathcal{M}_1)$

(b) $(\mathcal{S}_1', \mathcal{M}_1')$

Fig. 6.3 Characteristic graphs for (**a**) $(\mathcal{S}_1, \mathcal{M}_1)$ and (**b**) $(\mathcal{S}_1', \mathcal{M}_1')$

(a) $(\mathcal{S}_1, \mathcal{M}_1)$

(b) $(\mathcal{S}_1', \mathcal{M}_1')$

Fig. 6.4 Results obtained in the attempt to maximize (6.43) over the pairs (p_0, p_1)

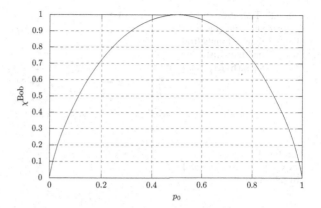

The quantum states in the DFS $\tilde{\mathcal{H}}_1$ are those from $\mathcal{S}_1'$. To obtain the secrecy capacity of this channel, the software Mathematica® was adopted in the attempt to obtain a maximum value for χ^{Bob}:

$$C_S(\mathcal{E}_1) = \chi^{\text{Bob}}$$
$$= \max_{\{P\}} S\left(p_0 \cdot \rho_0 + p_1 \cdot \rho_1\right). \tag{6.43}$$

To reach this objective, we used an exhaustive search among 30,000 pairs of (p_0, p_1) respecting the restriction that $p_0 + p_1 = 1$. The graphic shown in Fig. 6.4 is a result of such search. As it can be seen, the maximum value for Bob's Holevo quantity is 1. This result was already expected since equal probabilities maximize the von Neumann entropy (6.43).

This way, for the channel $\mathcal{E}_1$, the zero-error secrecy capacity is

$$C_S^{(0)}(\mathcal{E}_1) = \min\left\{C^{(0)}(\mathcal{E}_1), C_S(\mathcal{E}_1)\right\}$$
$$= \min\{1, 1\}$$
$$= 1 \text{ bits per symbol per channel use.}$$

It is possible to conclude, from this first example, that there are quantum channels $\mathcal{E}$ whose zero-error secrecy capacity is strictly positive, i.e., $C_S^{(0)}(\mathcal{E}) > 0$.

Example 6.4 (Non-Trivial ZESC). In this second example, the quantum channel $\mathcal{E}_2$ has positive zero-error capacity reached by an optimum pair $(\mathcal{S}_2, \mathcal{M}_2)$ where $\mathcal{S}_2 = \{\rho_1, \ldots, \rho_6\}$ and $\mathcal{M}_2 = \{M_i = |\rho_i\rangle \langle\rho_i|\}_{i=1}^6$. The model of errors for the channel is shown in Fig. 6.5a. Since we are interested in the adjacency relations, the probabilities were omitted.

From the pair $(\mathcal{S}_2, \mathcal{M}_2)$ we obtained the pair $(\mathcal{S}_2', \mathcal{M}_2')$ were $\mathcal{S}_2' = \{\rho_2, \rho_3, \rho_5\}$ and $\mathcal{M}_2' = \{M_2, M_3, M_5\}$. The relation between input and output states is depicted in Fig. 6.5b.

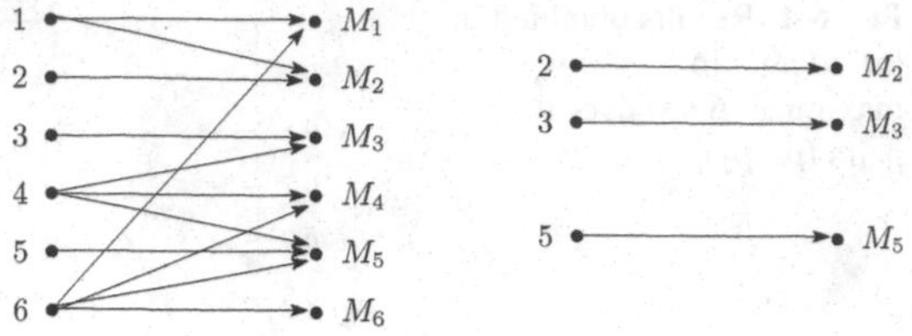

Fig. 6.5 Transitions performed by the quantum channel $\mathcal{E}_2$ over inputs from the optimum pairs (**a**) $(\mathcal{S}_2, \mathcal{M}_2)$ and (**b**) $(\mathcal{S}'_2, \mathcal{M}'_2)$

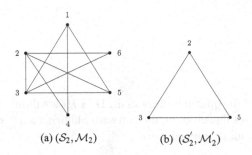

(a) $(\mathcal{S}_2, \mathcal{M}_2)$

(b) $(\mathcal{S}'_2, \mathcal{M}'_2)$

Fig. 6.6 Characteristic graphs of (**a**) $(\mathcal{S}_2, \mathcal{M}_2)$ and (**b**) $(\mathcal{S}'_2, \mathcal{M}'_2)$

(a) $(\mathcal{S}_2, \mathcal{M}_2)$

(b) $(\mathcal{S}'_2, \mathcal{M}'_2)$

Characteristic graphs for $\mathcal{E}_2$ with inputs $(\mathcal{S}_2, \mathcal{M}_2)$ and $(\mathcal{S}'_2, \mathcal{M}'_2)$ can be found in Fig. 6.6a, b, respectively. The clique number $\omega(\tilde{\mathcal{G}}(\mathcal{E}_2))$ is equal to 3 and can be obtained from the vertices $(2, 3, 5)$, $(1, 3, 5)$, or also $(2, 3, 6)$ considering the graph in Fig. 6.6a. On the other hand, the clique of the graph in Fig. 6.6b is also equal to 3, but obtained directly from the vertices $(2, 3, 5)$.

The quantum zero-error capacity of $\mathcal{E}_2$ considering the pair $(\mathcal{S}'_2, \mathcal{M}'_2)$ is

$$C^{(0)}(\mathcal{E}_2) = \sup_{\tilde{\mathcal{H}}_2} \sup_n \frac{1}{n} \log \dim(\tilde{\mathcal{H}}_2)^n$$

$$= \log 3$$

$$\approx 1,5849 \text{ bits per symbol per channel use.} \tag{6.44}$$

Aiming at quantifying $C_S(\mathcal{E}_2)$, Bob's Holevo quantity (6.45) was obtained with the software Mathematica® in the attempt to maximize it over the triple (p_1, p_2, p_3) under the restriction $p_1 + p_2 + p_3 = 1$.

$$C_S(\mathcal{E}_2) = \chi^{\text{Bob}} = \max_{\{P\}} S(p_1 \cdot p_2 + p_2 \cdot p_3 + p_3 \cdot p_5). \tag{6.45}$$

The exhaustive search considered 20,000 valid triples. The results obtained are presented in Fig. 6.7, which shows the graphic obtained in two different perspectives. According to the results observed, the highest value observed for χ^{Bob} was 1.5849 bits per symbol per channel use.

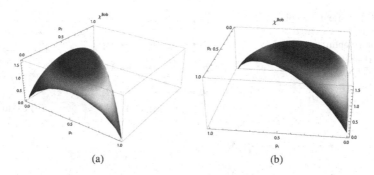

Fig. 6.7 Two different perspectives for the graph of Holevo quantity (6.45) with exhaustive search over the pairs (p_1, p_2, p_3). **(a)** Perspective 1 **(b)** Perspective 2

With these results, we have that the zero-error secrecy capacity of $\mathcal{E}_2$ is

$$C_S^{(0)}(\mathcal{E}_2) \geq \min\left\{C^{(0)}(\mathcal{E}_2), C_S(\mathcal{E}_2)\right\}$$

$$\geq \min\{1.5849, 1.5849\}$$

$$\geq 1.5849 \text{ bits per symbol per channel use.}$$

From this example, we can conclude that there are quantum channels $\mathcal{E}$ whose zero-error secrecy capacity is non-trivial, i.e., $C_S^{(0)}(\mathcal{E}) > 1$. We cannot guarantee that the ZESC of $\mathcal{E}_2$ is 1.5849 because we considered the case for $n = 1$. We do not have knowledge if there exists other DFS with higher dimensions for different values of n.

The collective amplitude damping quantum channel [1] has ZESC equal to the one of $\mathcal{E}_2$, characterizing a practical example of the non-triviality of this capacity.

The equality between the zero-error and secrecy capacities verified in the results of the quantum channel $\mathcal{E}_2$ is not a surprise. It happens because it is possible to derive an optimum pair $(\mathcal{S}_2', \mathcal{M}_2')$ from $(\mathcal{S}_2, \mathcal{M}_2)$. This example illustrates a quantum channel which is in the first situation of Theorem 6.5.

Example 6.5 (Situation 2 of Theorem 6.5). In the examples shown previously, we have that $C^{(0)}(\mathcal{E}) = C_S(\mathcal{E})$, emphasizing occurrences of the first situation in the proof of Theorem 6.5. The third example illustrates the second situation described.

Let $\mathcal{E}_3$ be a quantum channel whose model of errors is composed by four elements: $E_0 = |0\rangle\langle 0|, E_1 = |1\rangle\langle 1|, E_2 = \frac{1}{2}|2\rangle\langle 2| + \frac{1}{2}|3\rangle\langle 2|$, and $E_3 = \frac{1}{2}|3\rangle\langle 3| + \frac{1}{2}|2\rangle\langle 3|$, i.e., $\mathcal{E}_3 \equiv \{E_i\}_{i=0}^{3}$. We have that $\mathcal{S}_3 = \{\rho_i = |i\rangle\langle i|, i = 0, \ldots, 3\}$. The mappings of the channel $\mathcal{E}_3$ over the inputs from $(\mathcal{S}_3, \mathcal{M}_3)$ are shown in Fig. 6.8a.

Upon considering the channel $\mathcal{E}_3$, we can see that its quantum zero-error capacity is equal to $C^{(0)}(\mathcal{E}_3) = \log 3$ bits per symbol per channel use, considering three classical messages associated with the input states in the following way: $0 \mapsto \rho_0$, $1 \mapsto \rho_1, 2 \mapsto \rho_2$, and $2 \mapsto \rho_3$. However, in the attempt to obtain the quantum secrecy capacity of $\mathcal{E}_3$, it is not possible to obtain a pair $(\mathcal{S}_3', \mathcal{M}_3')$ which is also

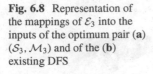

Fig. 6.8 Representation of the mappings of $\mathcal{E}_3$ into the inputs of the optimum pair **(a)** $(\mathcal{S}_3, \mathcal{M}_3)$ and of the **(b)** existing DFS

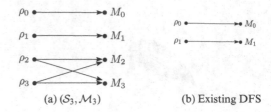

(a) $(\mathcal{S}_3, \mathcal{M}_3)$ · (b) Existing DFS

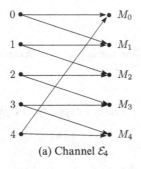

(a) Channel $\mathcal{E}_4$

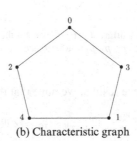

(b) Characteristic graph

Fig. 6.9 Quantum **(a)** channel $\mathcal{E}_4$ and its **(b)** characteristic graph

optimum, because the transitions that cause $\mathcal{E}_3(\rho_2) = \rho_3$ and $\mathcal{E}_3(\rho_3) = \rho_2$ result in an interaction with the environment. Such interaction causes information leakage which is not adequate for a quantum secrecy scenario. However, this channel has a DFS with 2 states, ρ_0 and ρ_1, shown in Fig. 6.8b.

This way, the ZESC of $\mathcal{E}_3$ is

$$C_S^{(0)}(\mathcal{E}_3) = \min \left\{ C^{(0)}(\mathcal{E}_3), C_S(\mathcal{E}_3) \right\}$$

$$= \min \{\log 3, \log 2\}$$

$$= 1 \text{ bits per symbol per channel use.}$$

Example 6.6 (Quantum Channel With No Zero-Error Secrecy Capacity). In the previous examples we saw that $C_S^{(0)}(\mathcal{E}) \neq 0$, but it is important to show that it is not always true. For the quantum channel $\mathcal{E}_4$ in Fig. 6.9a, whose characteristic graph is shown in Fig. 6.9b, we have that $C^{(0)}(\mathcal{E})$ is reached by an optimum pair $(\mathcal{S}_4, \mathcal{M}_4)$, with $\mathcal{S}_4 = \{ |00\rangle, |12\rangle, |24\rangle, |31\rangle, |43\rangle \}$ and $\mathcal{M}_4 = \{M_{0,0}, M_{1,2}, M_{2,4}, M_{3,1}, M_{4,3}\}$, where $\sum_{M \in \mathcal{M}_4}^{4} M \leq \mathbb{1}$.

Finding the zero-error capacity of the classical channel corresponding to $\mathcal{E}_4$ was a problem proposed by Shannon [49] whose solution was presented 20 years later by Lovász [36]. In the quantum case, the quantum zero-error capacity is reached after two or more uses of the channel, as shown by Medeiros [38, p. 70]. Such result was previously shown in Example 5.4.

The channel $\mathcal{E}_4$ is not unital and there is no $M_i \in \mathcal{M}$ that satisfies the condition $\mathcal{E}(M_i) = M_i\mathcal{E}M_i$. This way, there is no DFS in the inner structure of the error-free code associated with the channel. It means that every input performs a transition which causes an undesired interaction with the environment which can lead to an information leakage. This way, we have that the quantum secrecy capacity of $\mathcal{E}_4$ is

$$C_S^{(0)} = \min \left\{ C^{(0)}(\mathcal{E}), C_S(\mathcal{E}) \right\}$$
$$= \min \left\{ \frac{1}{2} \log 5, 0 \right\}$$
$$= 0.$$

This example illustrates that despite some channels have positive and non-trivial quantum zero-error capacity obtained from two or more uses of the channel, the nonexistence of a DFS causes $\mathcal{M}' = \varnothing$. It results that no pair $(\mathcal{S}', \mathcal{M}')$ can be used to encode messages without decoding errors and with secrecy. In other words, $C_S^{(0)}(\mathcal{E}_4) = 0$.

6.7 Related Literature

Until the first articles describing the results presented in Sect. 6.3 [20–23], many works in the literature explored the use of DFS in communication, but not considering their capability to send unconditionally secure messages. Among some works it was possible to see applications of DFS in protocols for quantum secure direct communication and for quantum deterministic secure communication [3, 12, 43]. In such protocols there is redundancy and eavesdropping check which increases significantly the number of messages exchanged in order to perform the communication with security. By using results previously discussed [23], all these protocols could be simplified with less message exchanges but with the same security, as can be seen in more detail in [19].

Regarding quantum wiretap channels, a few codes for this purpose were found, as presented previously in Sect. 6.2. The codes proposed by Hamada [25, 26] are based on CSS codes and, according to the author, can be easily used for practical implementation since they do not demand resources as entanglement. However, the rate of these codes is below the quantum secrecy capacity of the channel. The work of Wilde, Guha, and Dutton [16, 53] shows a code for quantum wiretap channels based on polar codes. The authors discuss that these codes can be restricted to certain quantum channels. Regarding the proposition of wiretap codes from DFS and quantum error-avoiding codes, as shown in Sect. 6.3, no similar strategies were found so far.

Braunstein et al. [4] enlighten the relation between DFS and zero-error subspaces, showing how the last is an instance of the former. Besides, the authors also proposed a method to find DFS in zero-error subspaces which has similarities with the method of Medeiros et al. [39]. In the characterization of the zero-error secrecy capacity we opted out to use the method of Medeiros et al. because it is guaranteed optimum and because it helped in showing a more intuitive approach to find a DFS in a quantum channel with positive zero-error capacity. It is important to emphasize that the work of Braunstein et al. [4] has other results, such as lower and upper bounds for the dimension of such subspaces.

Starting from confusability graphs for quantum channels, the work of Chiribella and Yang [8] aims at searching for connected components to identify, among others, decoherence-free subspaces. The work of these authors, however, focus on quantum covariant channels and they did not explore the quantum zero-error capacity of such channels nor the relation with the confusability graphs considered by Duan et al. [15].

Regarding capacity, Watanabe [52] characterizes a class of *quantum channels more capable than the environment*. In these channels, the quantum capacity and the secrecy capacity are equal. However, the author shows that the conditions that make a channel of such kind are, in general, hard to verify.

6.8   Further Reading

This chapter aimed at showing the zero-error secrecy capacity, the highest rate according to which it is possible to exchange messages through certain noisy and wiretapped quantum channels without decoding errors nor information leakage. This capacity puts together concepts of quantum zero-error information theory, of quantum secrecy capacity, and of decoherence-free subspaces and subsystems. The results, when possible, were also shown in terms of graph theory and the security analysis was discussed. Detailed examples illustrated the concepts introduced. Relations with other works in literature were also presented.

The articles that introduce the concepts to build up the zero-error secrecy capacity can be found in [20–23]. The thesis in which the concept was fully characterized was published only in Portuguese [18].

Besides the quantum zero-error information theory, the other building blocks of ZESC which are the decoherence-free subspaces and quantum wiretap channels, covered in the sections of this chapter, are very interesting with many applications and with perspective for many developments. As a suggestion regarding DFS, we recommend the work of Lidar and Whaley [34] and the thesis of Bacon [1]. Regarding quantum wiretap channels, we recommend the seminal papers of Cai et al. [6] and Devetak [11]. The book of Hayashi [27, Sect. 9.5] contains a section regarding this subject in the context of discussing quantum communications over eavesdropped channels.

Regarding future work with ZESC, Shabani et al. [46, 47] discuss the existence of "more relaxed" conditions for the existence of DFS. Taking this into account, could such conditions be considered and implemented in practical scenarios to favor the positivity of the zero-error secrecy capacity in a more wide number of noisy quantum channels? Such answer could favor more practical implementations of quantum communications which are simultaneously error-free and secure.

References

1. Bacon DM (2001) Decoherence, control, and symmetry in quantum computers. Ph.D Thesis, University of California at Berkeley, USA
2. Beige A, Braun D, Tregenna B, Knight PL (2000) Quantum computing using dissipation to remain in a decoherence-free subspace. Phys Rev Lett 85:1762, doi:10.1103/Phys-RevLett.85.1762
3. Bin G, ShiXin P, Biao S, Kun Z (2009) Deterministic secure quantum communication over a collective-noise channel. Science in China Series G: Sci China Ser G Phys Mech Astron 52(12):1913–1918. doi:10.1007/s11433-009-0303-y
4. Braunstein SL, Kribs DW, Patra MK (2011) Zero-error subspaces of quantum channels. In: IEEE international symposium on information theory, Russia, pp 104–108
5. Byrd MS, Wu LA, Lidar DA (2004) Overview of quantum error prevention and leakage elimination. J Mod Opt 51(16–18):2449–2460. doi:10.1080/09500340408231803
6. Cai N, Winter A, Yeung RW (2004) Quantum privacy and quantum wiretap channels. Probl Inf Transm 40:318–336
7. Casati GBG, Strini G (2007) Principles of quantum computation and information. In: Basic tools and special topics, vol II. World Scientific, Singapore
8. Chiribella G, Yang Y (2013) Confusability graphs for symmetric sets of quantum states. In: XXIX international colloquium on group-theoretical methods in physics, World Scientific, Tianjin, China, pp 251–256
9. Choi MD, Kribs DW (2006) A method to find quantum noiseless subsystems. Phys Rev Lett 96:501–506
10. Davidson K (1996) C^*-algebras by example. Fields institute monographs. American Mathematical Society, Rhode Island
11. Devetak I (2005) The private classical capacity and quantum capacity of a quantum channel. IEEE Trans Inf Theory 51(1):44–55
12. Dong HK, Dong L, Xiu XM, Gao YJ (2010) A deterministic secure quantum communication protocol through a collective rotation noise channel. Int J Quantum Inf 8(8):1389–1395
13. Dorner U, Klein A, Jaksch D (2008) A quantum repeater based on decoherence free subspaces. Quantum Inf Comput 8:468
14. Duan LM, Guo GC (1999) Quantum error avoiding codes versus quantum error correcting codes. Phys Lett A 255:209–212. doi:10.1016/S0375-9601(99)00183-8
15. Duan R, Severini S, Winter A (2013) Zero-error communication via quantum channels, non-commutative graphs and a quantum Lovasz ϑ function. IEEE Trans Inf Theory 59(2):1164–1174
16. Dutton Z, Guha S, Wilde MM (2012) Performance of polar codes for quantum and private classical communication. In: 50th annual allerton conference on communication, control, and computing, Illinois, pp 1–8
17. Feng M (2001) Quantum computing and communication with decoherence-free atomic states. http://arxiv.org/abs/quant-ph/0111041. Accessed 2 Dec 2012
18. Guedes EB (2013) Capacidade quântica de sigilo erro-zero e informação acessível erro-zero de fontes quânticas. Ph.D Thesis, Universidade Federal de Campina Grande, Brazil
19. Guedes EB, de Assis FM (2012) Enhancing quantum protocols with the security of decoherence-free subspaces and subsystems. In: IV workshop school of quantum computation and information, Fortaleza, Brazil, pp 1–8

20. Guedes EB, de Assis FM (2012) Quantum zero-error secrecy capacity. In: IV workshop school of quantum computation and information, Fortaleza, Brazil, pp 1–8
21. Guedes EB, de Assis FM (2012) Unconditional security with decoherence-free subspaces. http://arxiv.org/abs/1204.3000. Accessed 30 Mar 2012
22. Guedes EB, de Assis FM (2012) Utilização de subespaços livres de descoerência em comunicações quânticas incondicionalmente seguras. In: Simpósio Brasileiro de Telecomunicações – SBrT'12, Brasília, Brazil, pp 1–5
23. Guedes EB, de Assis FM (2013) On the security of decoherence-free subspaces and subsystems for classical information conveying through quantum channels. Int J Quantum Inf 11(2):1350022-1–1350022-14
24. Gyongyosi L, Imre S (2012) Long-distance quantum communications with superactivated gaussian optical quantum channels. Opt Eng 51(1):1–16
25. Hamada M (2008) Algebraic and quantum theoretical approach to coding on wiretap channels. In: International symposium on communications, control and signal processing, Malta, pp 1–6
26. Hamada M (2008) Constructive codes for classical and quantum wiretap channels. Nova Science Publishers, New York, pp 1–48
27. Hayashi M (2006) Quantum Information – An Introduction. Springer, Japan
28. Holevo AS (1998) The capacity of the quantum channel with general signal states. IEEE Trans Inform Theory 4(1):269–273
29. Ivanov PA, Poschinger UG, Singer K, Schmidt-Kaler F (2010) Quantum gate in the decoherence-free subspace of trapped-ion qubits. Europhysics Letters 92(3):30006
30. Jaeger G, Sergienko A (2008) Constructing four-photon states for quantum communication and information processing. Int J Theoret Phys 47:2120
31. Kielpinski D (2001) A decoherence-free quantum memory using trapped ions. Science 291:1013
32. Knill E, Laflamme R, Viola L (2000) Theory of quantum error correction for general noise. Phys Rev Lett 84:2525
33. Kwiat PG, Berglund AJ, Altepeter JB, White AG (2000) Experimental verification of decoherence-free subspaces. Science 290:498–501
34. Lidar DA, Whaley KB (2003) Decoherence-Free Subspaces and Subsystems, Springer Lecture Notes in Physics, Berlin, pp 83–120
35. Lidar DA, Chuang IL, Whaley KB (1998) Decoherence-free subspaces for quantum computation. Phys Rev Lett 81:2594–2597
36. Lovász L (1979) On the Shannon capacity of a graph. IEEE Trans Inform Theory 25(1):1–7
37. Mayers D (2001) Unconditional security in quantum cryptography. J ACM 48(3):351–406
38. Medeiros RAC (2008) Zero-error capacity of quantum channels. PhD thesis, Universidade Federal de Campina Grande – TELECOM Paris Tech
39. Medeiros RA, Alleaume R, Cohen G, de Assis FM (2006) Zero-error capacity of quantum channels and noiseless subsystems. In: IEEE International Telecommunications Symposium, Fortaleza, Brazil, pp 900–905, 3–6 Sept 2006
40. Medeiros RAC, Alleaume R, Cohen G, de Assis FM (2006) Quantum states characterization for the zero-error capacity. http://arxiv.org/abs/quant-ph/0611042, accessed 25 Oct. 2013
41. Mohseni M, Lundeen JS, Resch KJ, Steinberg AM (2003) Experimental application of decoherence-free subspaces in an optical quantum-computing algorithm. Phys Rev Lett 91:187903
42. Nielsen MA, Chuang IL (2010) Quantum Computation and Quantum Information. Cambridge University Press, Cambridge, England
43. Qin S, Wen Q, Meng L, Zhu F (2009) Quantum secure direct communication over the collective amplitude damping channel. Science in China Series G: Physics, Mechanics and Astronomy 52(8):1208–1212, DOI 10.1007/s11433-009-0140-z
44. Schumacher B, Westmoreland MD (1997) Sending classical information via noisy quantum channels. Phys Rev A 56:131–138, doi: 10.1103/PhysRevA.56.131
45. Schumacher B, Westmoreland M (1998) Quantum privacy and quantum coherence. Phys Rev Lett 80(25):5695–5697

46. Shabani A (2009) Open quantum systems and error correction. Ph.D Thesis, University of Southern California
47. Shabani A, Lidar DA (2005) Theory of initialization-free decoherence-free subspaces and subsystems. Phys Rev A 72:042303. doi:10.1103/PhysRevA.72.042303
48. Shannon CE (1949) Communication theory of secrecy systems. Bell Syst Tech J 28(4): 656–715
49. Shannon CE (1956) The zero error capacity of a noisy channel. IRE Trans Inf Theory 2(3):8–19
50. Subramaniany A, Suresh AT, Raj S, Thangaraj A, Blochy M, McLaughliny S (2010) Strong and weak secrecy in wiretap channels. In: International symposium on turbo codes and iterative information processing, Brest, France, pp 30–34
51. Viola L, Fortunato EM, Pravia MA, Knill E, Laflamme R, Cory DG (2001) Experimental realization of noiseless subsystems for quantum information processing. Science 293:2059–2063. doi:10.1103/PhysRevA.85.012326
52. Watanabe S (2012) Private and quantum capacities of more capable and less noisy quantum channels. Phys Rev A 85:012326. doi:10.1103/PhysRevA.85.012326
53. Wilde MM, Guha S (2011) Polar codes for degradable quantum channels. http://arxiv.org/abs/1109.5346. Accessed 13 Feb 2016
54. Wyner AD (1975) The wire-tap channel. Bell Syst Tech J 54(8):1355–1387
55. Xia Y, Song J, Yang ZB, Zheng SB (2010) Generation of four-photon polarization-entangled decoherence-free states within a network. Appl Phys B 99:651–656
56. Xue P (2008) Long-distance quantum communication in a decoherence-free subspace. Phys Lett A 372:6859–6866
57. Xue P, Xiao YF (2006) Universal quantum computation in decoherence-free subspace with neutral atoms. Phys Rev Lett 97:140501
58. Zanardi P, Rasetti M (1997) Noiseless quantum codes. Phys Rev Lett 79:3306. doi:10.1103/PhysRevLett.79.3306
59. Zhang XD, Zhang Q, Wang ZD (2006) Physical implementation of holonomic quantum computation in decoherence-free subspaces with trapped ions. Phys Rev A 74:034302

Chapter 7
Zero-Error Accessible Information of a Quantum Source

A *quantum source* is an essential component of quantum communication system because it corresponds to the set of quantum symbols that will be used to encode classical messages. In this encoding process, there is a bijective mapping between messages and quantum states, but each quantum state is associated with a certain probability. Differently from classical messages that are completely distinguishable, quantum states may not necessarily be so. A consequence is that the classical information encoded by the quantum source may not be fully recoverable after a measurement.

Considering this intrinsic difficulty to recover information from quantum sources, an information measure, called *accessible information*, has been proposed in the literature [12, Sect. 12.1]. It establishes the maximum amount of classical information that can be retrieved after being encoded by a quantum source. Because all measurement strategies can be used to retrieve information from the quantum system, calculating the accessible information is a hard task in general. Fortunately, there are some useful upper and lower bounds that are easiest to calculate and that give good estimates for the accessible information [3, 4, 7, 8, 16].

Aiming at avoiding errors in decoding messages from quantum sources, this chapter presents some recent results regarding an information measure for quantum sources, called *Zero-Error Accessible Information* (ZEAI). This quantity represents the maximum amounts of bits per symbol that can be retrieved from a quantum source with no decoding errors. The ZEAI of a quantum source unifies concepts from quantum sources, accessible information, classical zero-error information theory and also from graph theory.

To introduce these results, this chapter is organized as follows. In Sect. 7.1 we revisit some fundamental concepts such as the formal definition of a memoryless quantum source, its entropy, accessible information, and the Holevo bound, which is an upper bound for accessible information. Section 7.2 introduces the ZEAI of a quantum source and its relation with classical zero-error channels. The relation

© Springer International Publishing Switzerland 2016
E.B. Guedes et al., *Quantum Zero-Error Information Theory*,
DOI 10.1007/978-3-319-42794-2_7

between ZEAI and graph theory is elucidated in Sect. 7.3. After that, some detailed examples are given in Sect. 7.4. The relation of ZEAI and other works in the literature is described in Sect. 7.5.

7.1 Accessible Information of Quantum Sources

To study accessible information in the quantum information theory domain, we take into account the canonical communication scheme shown in Fig. 7.1. The quantum source encodes classical messages in quantum states as described in Definition 7.1.

Definition 7.1 (Memoryless Quantum Source). Let $\mathcal{A} = \{0, \ldots, \ell\}$ be a set of classical messages. A memoryless quantum source is a device that prepares quantum states according to an ensemble $\{\rho_i, p_i\}$. The set $\mathcal{S} = \{\rho_0, \ldots, \rho_\ell\}$ is denoted the *source alphabet*, commonly composed of pure non-orthogonal quantum states ρ_i, called *quantum letters*. The quantum letter ρ_i is associated with the classical message i. The quantum source outputs a letter ρ_i with probability p_i, where $\sum_{i=0}^{\ell} p_i = 1$. For a given sequence a of classical messages, $a = a_1 a_2 \ldots a_n, a \in \mathcal{A}^n$, the corresponding quantum state prepared by the quantum source, called *quantum codeword*, is given by the tensor product of the corresponding quantum letters, i.e.,

$$\rho(a) = \rho_{a_1} \otimes \rho_{a_2} \otimes \ldots \otimes \rho_{a_n}. \tag{7.1}$$

According to Definition 2.7, the ensemble $\{\rho_i, p_i\}$ of a quantum source can also be represented by the corresponding density operator

$$\rho = \sum_{i=0}^{\ell} p_i \rho_i. \tag{7.2}$$

Taking into account such density operator for a quantum source, we can introduce the *entropy of a quantum source*.

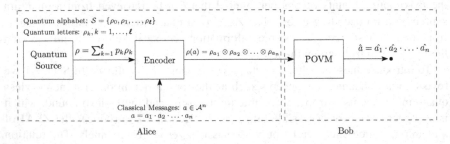

Fig. 7.1 Canonical communication model in which there is a single sender and a single receiver

Definition 7.2 (Entropy of a Quantum Source). The entropy of a quantum source is the von Neumann entropy of the density operator (7.2) that describes the ensemble $\{\rho_i, p_i\}$, i.e.,

$$S(\rho) = -\operatorname{Tr} \rho \log \rho. \tag{7.3}$$

We assume that Alice has a quantum source with the given description and that she prepares the quantum state ρ. Alice gives such quantum state to Bob, who can adopt a POVM measurement scheme aiming at identifying the corresponding message sent by Alice. The measurement outputs are arguments for a decoding function. The decoder must decide which classical message was originally sent by Alice.

A quantum source is *purely classical* if the corresponding source alphabet S is composed of pairwise orthogonal quantum states, since such states are completely distinguishable at the receiver's end. If the set S contains nonorthogonal quantum states, then there is no measurement strategy that can extract all the information about the quantum source. A third situation considers that the states of the quantum source are nonorthogonal but with commuting density matrices. In this case we have a *broadcast source* in which given two quantum systems that are not a copy of the source, their partial trace results in the state of the quantum source [2].

We recall the accessible information, presented in Definition 3.18. It is a measure of how well one can infer the state prepared by the source by measuring its output. In quantum information theory, there is no general method to calculate the accessible information of a quantum source. However, several lower and upper bounds for the accessible information were demonstrated; the most important is the *Holevo bound* [12, Chap. 12].

Theorem 7.1 (Holevo Bound [7]). *Suppose that Alice has a quantum source F with ensemble $\{\rho_i, p_i\}$. She sends Bob some quantum letters emitted by F. Bob measures the received quantum letters with a POVM $\{M_i\}_{i=0}^{m}$ and obtains B. The Holevo bound states that for any measurement scheme used by Bob, we have that*

$$I_{acc}(F) \leq S(\rho) - \sum_{k=0}^{\ell} p_k S(\rho_k), \tag{7.4}$$

where ρ is the density operator given by (7.2).

The right side of (7.4) is known as the Holevo quantity and it is denoted by χ. Taking into account the concavity of the entropy, we have $I(A; B) \leq \chi \leq H(A)$.

Example 7.1 (Holevo Bound). Suppose that Alice has a quantum source F that emits the states ρ_0 and ρ_1 according to a uniform distribution, where $|\psi_0\rangle = |0\rangle$, $|\psi_1\rangle = \cos\theta\,|0\rangle + \sin\theta\,|1\rangle$, $\rho_0 = |\psi_0\rangle\langle\psi_0|$, and $\rho_1 = |\psi_1\rangle\langle\psi_1|$. From Example 3.17, we know that $S(\rho) = H(\frac{1}{2} - \frac{\cos\theta}{2})$. Remember that ρ_0 and ρ_1 are pure states. Again, consider that A is the index of the state emitted by the source and B corresponds to the measurement result. The best measurement strategy to discriminate between ρ_0 and ρ_1 is given by the projective measurement $\mathcal{M} = \{E_0 = |w_0\rangle \langle w_0|, E_1 = |w_1\rangle \langle w_1|\}$, where

$$|w_0\rangle = \cos\left(\frac{\pi}{4} + \frac{\theta}{2}\right)|0\rangle + \sin\left(\frac{\pi}{4} + \frac{\theta}{2}\right)|1\rangle, \tag{7.5}$$

$$|w_1\rangle = \cos\left(-\frac{\pi}{4} + \frac{\theta}{2}\right)|0\rangle + \sin\left(-\frac{\pi}{4} + \frac{\theta}{2}\right)|1\rangle. \tag{7.6}$$

Notice that $\langle w_0| w_1\rangle = 0$ and that $E_0 + E_1 = \mathbb{1}$. The conditional probability that $B = b$ given that $A = a$, for $a, b \in \{0, 1\}$, is calculated as

$$p(b|a) = \text{Tr}(E_b |\psi_a\rangle \langle \psi_a|) = |\langle w_b| \psi_a\rangle|^2 = \begin{cases} \frac{1+\sin\theta}{2} & \text{if } a = b \\ \frac{1-\sin\theta}{2} & \text{if } a \neq b \end{cases} \tag{7.7}$$

It is possible to see that we can make an analogy with a binary symmetric channel, where A is the input, B is the output, and the error probability is $\frac{1-\sin\theta}{2}$ (see Example 3.8). The random variable A is uniformly distributed at the channel input and the random variable B is also uniformly distributed. Then, $H(A) = H(B) = 1$. The mutual information is given by

$$I(X; Y) = H(Y) - H(Y|X)$$

$$= 1 - H\left(\frac{1 - \sin\theta}{2}\right). \tag{7.8}$$

Since the measurement scheme is proven to be optimal [1], we have that $I_{\text{acc}}(F) = 1 - H\left(\frac{1-\sin\theta}{2}\right)$. Figure 7.2 shows plots of accessible information $I_{\text{acc}}(F)$ and Holevo quantity $\chi(F)$ versus the parameter θ. For any $0 \leq \theta \leq \pi$, we see that $I_{\text{acc}}(F) \leq \chi(F)$, with equality when $\theta = \pi/2$, i.e., when the states ρ_0 and ρ_1 are completely distinguishable.

Fig. 7.2 Plots of accessible
information and Holevo
bound for a quantum source F

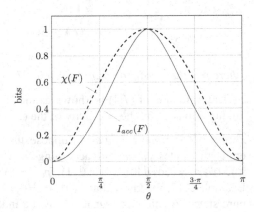

In the previous example, the parameter $\theta = \pi/2$ led the quantum letters to be orthogonal and the accessible information was maximum. Unfortunately, there are situations where Bob is not able to infer the state given by Alice.

Example 7.2 (Measurement with Inconclusive Results). Suppose that a quantum source can produce two quantum states $|\psi_1\rangle = |0\rangle$ or $|\psi_2\rangle = \frac{|0\rangle + |1\rangle}{\sqrt{2}}$ with equal probability. Since $\langle\psi_1|\psi_2\rangle \neq 0$, it is not possible to precisely determine what state was emitted by the source. However, it is possible to perform measurements that can distinguish these states most of the time. To do this, define a POVM with the following three elements:

$$E_1 = \frac{\sqrt{2}}{1+\sqrt{2}}\,|1\rangle\langle1|,$$

$$E_2 = \frac{\sqrt{2}}{1+\sqrt{2}}\,\frac{(|0\rangle - |1\rangle)(\langle0| - \langle1|)}{2},$$

$$E_3 = \mathbb{1} - E_1 - E_2.$$

Suppose that the state $|\psi_1\rangle$ was delivered by the quantum source. After performing a measurement with POVM operators $\{E_1, E_2, E_3\}$, the probability of getting 1 is zero, since $\langle\psi_1|E_1|\psi_1\rangle = 0$. Therefore, if measurement outcomes 1, then we can conclude with certainty that the state emitted by the source was $|\psi_2\rangle$. In a similar way, if the measurement result is 2, then we conclude that the source outputted the state $|\psi_1\rangle$. Certain times, however, the result will be 3 and the measurement is inconclusive. In summary, every time we get outputs 1 and 2 we can infer the quantum state emitted by the source without confusion. Otherwise, we can infer nothing, since $p_1 = p_2 = 1/2$.

7.2 Zero-Error Accessible Information of a Quantum Source

We are going to consider the simplified communication scheme of Fig. 7.1. Alice has a quantum source that emits quantum letters to be measured by Bob using a POVM. When Bob attempts to identify the received message, we assume that *no decoding errors can occur*. In other words, by using measurements, Bob must be 100 % sure about the original message sent by Alice. Next, we define the *Zero-Error Accessible Information* (ZEAI) of a quantum source.

Definition 7.3 (Zero-Error Accessible Information of a Quantum Source). Let A be a discrete random variable corresponding to message indexes that are associated with quantum letters emitted by a memoryless quantum source F. Now let B be a discrete random variable corresponding to measurement results of quantum states sent by the source. The zero-error accessible information of such quantum source, denoted by $I_{acc}^{(0)}(F)$, is the maximum amount of messages of length n sent by the source such that $H(A|B) = 0$.

In the original definition of accessible information, vide Definition 3.18, we have that the maximum is taken over the mutual information of the random variables A and B. This turn, in Definition 7.3, we have that the maximum is taken over the set of messages that makes the conditional entropy of A and B equal to zero.

The equality in $H(A|B) = 0$ of Definition 7.3 means that the uncertainty of the random variable A is zero when the random variable B is known, showing the zero-error behavior in this measure of information.

The next step is to obtain a numerical expression for the ZEAI of a quantum source.

Theorem 7.2 (Numerical Expression for ZEAI). *Let $N(n)$ be the set of quantum codewords of length n which can be sent by a memoryless quantum source F and that can be retrieved with no error with a POVM measurement. The zero-error accessible information of F is given by*

$$I_{acc}^{(0)}(F) \triangleq \sup_{n \to \infty} \frac{1}{n} \log N(n). \tag{7.9}$$

Proof. To demonstrate the theorem, we consider that both the emission of quantum letters by the quantum source and the subsequent measurement with no decoding errors are equivalent to the zero-error capacity of a discrete memoryless quantum channel.

To do so, we consider that the state produced by the quantum source and the output of the POVM can be written as a discrete memoryless classical channel W : $A \to B$ with the following stochastic matrix

$$W(a,b) \triangleq \Pr[B = b|A = a] = \mathrm{Tr}(\rho_a M_b), (a,b) \in \mathcal{A} \times \mathcal{B}, \tag{7.10}$$

where ρ_a is the quantum letter emitted by the source; M_b is the operation element of the POVM used for measurement; and $\mathcal{A}$ and $\mathcal{B}$ are the alphabets of the random variables A and B, respectively. When the source emits k quantum letters, we have

$$W^k(a^k, b^k) = \prod_{i=1}^{k} W(a_i, b_i). \tag{7.11}$$

Considering this interpretation, the maximum amount of messages per number of source emissions that can be sent over the channel W with no decoding errors is corresponding to its zero-error capacity, i.e., $I_{acc}^{(0)}(F) = C_0(W) = \sup_{n \to \infty} \frac{1}{n} \log N(n)$. Thus, we conclude the proof.

Theorem 7.2 shows an interesting aspect: although zero-error accessible information is a measure of information that characterizes a quantum device, its computation depends on the classical zero-error capacity of a corresponding discrete memoryless channel. This is a counterintuitive result specially because there is no restriction or requirement regarding the quantum letters emitted by the source nor regarding their probabilities. Note that if the quantum source is purely classical, we have that all quantum letters are distinguishable. In this case, the zero-error accessible information is $\log |S|$, where S is the alphabet of the quantum source.

The accessible information of a classical source is not a relevant measure of information, since two classical states can always be distinguished. In contrast, the ZEAI of a quantum source is not trivial because quantum information cannot always be distinguished. It is important to emphasize that the definition of ZEAI of a quantum source imposes one restriction: *the absence of errors*. As a consequence, we can verify the following inequalities: $I_{\text{acc}}^{(0)}(F) \leq I_{\text{acc}}(F) \leq \chi(F)$ for a quantum source F.

7.3 Representation in Graphs

The zero-error capacity of classical channels has a formulation in terms of graph theory, as shown previously in Sect. 5.2. The problem of finding the zero-error accessible information of a quantum source is equivalent to the problem of obtaining the zero-error capacity of a classical channel. Concepts like orthogonality of input states and characteristic graphs of a quantum channel are straightforwardly defined for a quantum source.

Definition 7.4 (Orthogonality of Quantum Letters). Given two quantum letters $\rho_i = |\psi_i\rangle \langle\psi_i|$ and $\rho_j = |\psi_j\rangle\langle\psi_j|$ belonging to the alphabet S of a quantum source, we say that ρ_i and ρ_j are non-adjacent, denoted by $\rho_i \perp \rho_j$, if they are orthogonal, i.e., $\langle\psi_i| \psi_j\rangle = 0$.

Orthogonality of quantum letters can be extended to quantum codewords. Two quantum codewords $\rho(i) = \rho_{i_1} \otimes \rho_{i_2} \otimes \ldots \otimes \rho_{i_n}$ and $\rho(j) = \rho_{j_1} \otimes \rho_{j_2} \otimes \ldots \otimes \rho_{j_n}$ are said to be orthogonal or non-adjacent if there is at least one k, $1 \leq k \leq n$, such that the corresponding quantum letters ρ_{i_k} and ρ_{j_k} are non-adjacent. The characteristic graph can therefore be defined.

Definition 7.5 (Characteristic Graph of a Quantum Source). Let F be a memoryless quantum source according to Definition 7.1. The characteristic graph of F is given by $\mathcal{G}(F) = \langle V, E\rangle$, where the sets of vertices and edges are given as follows.

1. The vertex set V is given by the classical messages associated with quantum letters of the source alphabet S;
2. There is an edge connecting the vertices (i, j) if the corresponding quantum letters ρ_i and ρ_j, $i \neq j$, are non-adjacent.

The graph $\mathcal{G}(F)$ can be generalized for n quantum source outputs. The vertex set of $\mathcal{G}^n(F)$ is given by V^n, and the set of edges is composed by the pairs of vertices whose corresponding codewords of length n are orthogonal.

Vertices of the characteristic graph are connected if the corresponding quantum letters or codewords are fully distinguishable by means of a quantum measurement. A clique on the characteristic graph corresponds to a subset of quantum letters or codewords that are pairwise distinguishable. The zero-error accessible information can be defined in terms of the clique number of the characteristic graph.

Theorem 7.3 (ZEAI in terms of Graph Theory). *Let F be a quantum source with characteristic graph $\mathcal{G}(F)$ according to Definition 7.5. The accessible information of F is given by*

$$I_{acc}^{(0)}(F) = \sup_n \frac{1}{n} \log \omega(\mathcal{G}^n(F)), \tag{7.12}$$

where $\omega(\mathcal{G}^n(F))$ stands for the clique number of $\mathcal{G}^n(F)$.

Proof. As proved in Theorem 7.2, there is an equivalence between the zero-error accessible information of a quantum source F and the zero-error capacity of a discrete memoryless classical channel W. As we already know, two input letters are non-adjacent at the source output if the corresponding rows of the stochastic matrix $W(b|a)$ are orthogonal. It is straightforward to conclude that the characteristic graph of the quantum source is identical to the characteristic graph of the DMC W, whose zero error-capacity is given by the right side of (7.12).

Upon reducing the zero-error accessible information to the problem of identifying the clique number of a graph, we can see that the calculation of the ZEAI is an $\mathcal{NP}$-complete problem. The exponential hardness is due to the difficulties of identifying the best POVM for zero-error measurements as well as in determining the length n of quantum codewords that maximize the number of non-confusable codewords per source output.

7.4   Examples

This section shows how to obtain the zero-error accessible information for some quantum sources by following the procedures shown previously in Sect. 7.2.

Example 7.3 (Quantum Source With ZEAI Equal to Zero). Let F_1 be a quantum source that emits two states $\rho_1 = |0\rangle \langle 0|$ and $\rho_2 = |+\rangle \langle +| = \frac{1}{2}(|0\rangle \langle 0| + |0\rangle \langle 1| + |1\rangle \langle 0| + |1\rangle \langle 1|)$ according to the uniform distribution. The source ensemble is given by $\{\rho_i, p_i\}$, where $p_1 = p_2 = 0.5$. According to Definition 7.4, it is easy to verify that the quantum letters ρ_1 and ρ_2 are non-orthogonal, since $\langle 0|+\rangle = \frac{1}{\sqrt{2}}$. Thus, we have that $I_{acc}^{(0)}(F_1) = 0$.

Example 7.4 (ZEAI of a Purely Classical Quantum Source). Let F_2 be a quantum source that emits states belonging to the computational basis of the 8-dimensional Hilbert space, $\{\rho_0 = |0\rangle \langle 0|, \rho_1 = |1\rangle \langle 1|, \ldots, \rho_7 = |7\rangle \langle 7|\}$, each one with

Fig. 7.3 Characteristic graph
of the quantum source F_2

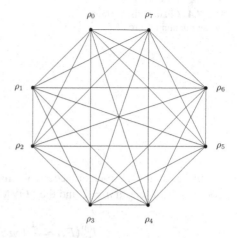

probability $1/8$. According to Definition 7.5, the characteristic graph of F_2 is shown
in Fig. 7.3.

Considering that states emitted by the source are pure and orthogonal, we have
that F_2 is a purely classical quantum source. A projective measurement scheme with
the POVM $\{M_{2,i} = |i\rangle\langle i|\}_{i=0}^{7}$ is sufficient to perfectly distinguish all the states. This
way, the zero-error accessible information of F_2 is

$$I_{\text{acc}}^{(0)}(F_2) = \sup_n \frac{1}{n} \log \omega(\mathcal{G}^n(F))$$

$$= \frac{1}{1} \log 8$$

$$= 3 \text{ bits per symbol.}$$

In this example, it is interesting to notice that individual measurements are
sufficient to reach the zero-error accessible information of this quantum source.
Because $\mathcal{G}(F)$ is a complete graph, $\omega(\mathcal{G}^n(F)) = \omega(\mathcal{G}(F))^n = |\mathcal{S}|^n$. Therefore,
$\frac{1}{n} \log |\mathcal{S}|^n = \log |\mathcal{S}|$.

Example 7.5 (ZEAI of a Quantum Source Corresponding to the Pentagon). Let F_3
be a quantum source that emits quantum letters of the alphabet $\mathcal{S} = \{\rho_0, \ldots, \rho_4\}$,
which is composed only of pure states that are not necessarily orthogonal. These
letters are emitted according to the uniform distribution. Figure 7.4 illustrates the
characteristic graph of F_3. This is the pentagon graph of the G_5 channel of Fig. 4.3,
whose zero-error capacity was calculated by Lovász and discussed in Sect. 4.3 and
in the Example 5.4.

Fig. 7.4 Characteristic graph
of the quantum source F_3

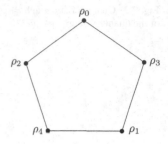

To reach the zero-error accessible information of F_3, we consider quantum codewords of length $n = 2$ and the POVM $\mathcal{M} = \{M_{00}, M_{12}, M_{24}, M_{31}, M_{43}\}$. Then,

$$I_{\text{acc}}^{(0)}(F_3) = \frac{1}{2} \log 5$$

$$\approx 1,1609 \text{ bits per symbol.} \qquad (7.13)$$

This example illustrates how collective measurements can extract more zero-error information from the quantum source than individual measurements. Moreover, we need more than one source emission in order to reach the zero-error accessible information.

7.5 Related Literature

The accessible information is not known for most quantum sources, and indeed, there is no general method for calculating this quantity. The Holevo bound is the famous and the most important upper bound of the accessible information [7]. The *Holevo χ quantity* is fundamental in proving several results in quantum information theory. Cerf and Adami [3] presented a formal proof for the bound and extended it to consider sequential measurements based on conditional and mutual entropies.

The first lower bound of the accessible information was conjectured by Wootters [16] and proved by Jozsa et al. [8]. Other bounds based on Jensen and Schwarz inequalities and also in purification schemes were also proposed in the literature. A survey can be found in Fuchs [4, Sect. 3.5].

Several numerical methods to calculate the accessible information were developed. In general, these methods seek for the best POVM that maximizes the mutual information $I(A; B)$. Nascimento and de Assis [10, 11] developed a method that is based on genetic algorithms. The open-source tool SOMIM (*Search for Optimal Measurements by an Iterative Method*) also considers this approach and makes use of an iterative method to find the best measurement scheme [9, 13, 15].

Sasaki et al. [14] addressed the problem of obtaining the accessible information of a quantum source, but restricted to the case of real and symmetric quantum sources. A quantum source is said to be real if the coefficients of the quantum letters are real numbers. The symmetry is verified when the states of the quantum source are equally spaced in a plane $x - z$ of the Bloch sphere. To calculate the accessible information of this kind of quantum source, the authors developed a method based on group theory to identify the best measurement. The resulting POVM has only three elements and can be implemented in a real scenario with existing technology, as discussed by the authors. Unfortunately, the results of Sasaki et al. [14] are restricted only to a certain class of quantum sources.

7.6 Further Reading

In this chapter we introduced the zero-error accessible information, a measure of information that gives the maximum amount of information that can be retrieved from a quantum source without errors. Obtaining the zero-error accessible information of a quantum source can be reduced to the problem of finding the zero-error capacity of an equivalent classical channel. The ZEAI involves concepts from quantum sources, accessible information, classical and quantum information theories and also from graph theory. Some examples illustrated the concepts and the relation with other existing works was discussed.

The zero-error accessible information was first proposed by Guedes and de Assis [6]. Later, on the thesis of Guedes [5], this concept was fully depicted. As discussed in Sect. 7.3, the problem of finding the ZEAI of a quantum source is equivalent to identifying the clique number of a graph, which is $\mathcal{NP}$-Complete.

Some authors developed heuristics to obtain the best POVM based on iterative methods [9, 13, 15]. Other strategies include genetic algorithms [10, 11].

References

1. Arikan E, Shumovsky A (2002) Holevo's bound. http://www.ee.bilkent.edu.tr/~qubit/n18holevo.ps. Accessed 28 May 2016
2. Bennett CH, Shor PW (1998) Quantum information theory. IEEE Trans Inf Theory 44(6):2724–2742
3. Cerf NJ, Adami C (1996) Accessible information in quantum measurement. http://arxiv.org/abs/quant-ph/9611032. Accessed 17 Feb 2014
4. Fuchs CA (1995) Distinguishability and accessible information in quantum theory. Ph.D Thesis, University of New Mexico, USA
5. Guedes EB (2013) Capacidade quântica de sigilo erro-zero e informação acessível erro-zero de fontes quânticas. Ph.D Thesis, Universidade Federal de Campina Grande, Brazil
6. Guedes EB, de Assis FM (2013) Informação acessível erro-zero de fontes quânticas. In: XXXI Simpósio Brasileiro de Telecomunicações, Fortaleza, Brazil, pp 1–5
7. Holevo AS (1973) Information theoretical aspects of quantum measurements. Probl Inf Transm 9(2):110–118

8. Jozsa R, Robb D, Wootters WK (1994) Lower bound for accessible information in quantum mechanics. Phys Rev A 49:668–677
9. Lee KL, Shang J, Chua WK, Looi SY, Englert BG (2011) SOMIM: an open-source program code for the numerical search for optimal measurements by an iterative method. http://www.quantumlah.org/publications/software/SOMIM/. Accessed 25 May 2013
10. Nascimento EJ, de Assis FM (2006) A numerical solution for the accessible quantum information problem. In: International telecommunications symposium, Fortaleza, Brazil, pp 495–500
11. Nascimento EJ, de Assis FM (2006) Soluções numéricas para o cálculo da informação acessível. In: Workshop-Escola de Computação e Informação Quânticas, Porto Alegre, Brazil, pp 265–274
12. Nielsen MA, Chuang IL (2010) Quantum computation and quantum information. Cambridge University Press, Cambridge
13. Rehacek J, Englert BG, Kaszlikowski D (2005) Iterative procedure for computing accessible information in quantum communication. Phys Rev A 71:054303
14. Sasaki M, Barnett SM, Jozsa R, Osaki M, Hirota O (1999) Accessible information and optimal strategies for real symmetrical quantum sources. Phys Rev A 59(5):3325–3335
15. Suzuki J, Assad SM, Englert BG (2007) Accessible information about quantum states: an open optimization problem. Chapman & Hall, Boca Raton, pp 309–348
16. Wootters WK (1992) Two extremes of information in quantum mechanics. In: Workshop on physics and computation, Dallas, Texas, pp 181–183

Chapter 8
Recent Developments in Quantum Zero-Error Information Theory

In the previous Chaps. 6–7 some recent developments and applications of the quantum zero-error information theory were introduced. In this chapter we introduce some contributions from other authors to the field.

This chapter is organized as follows. We revisit some nonlocal phenomena by Bell's inequalities and their consequences in Sect. 8.1. After that, we introduce some definitions that did not appear so far. We revisit relevant contrasts of classical and quantum correlations and discuss a proof of the Bell's inequality. Also, due to their importance, we revisit Gleason's and Kochen-Specker's theorems. We observe that this section has a historical flavor, so the reader, based on his own background, can skip this first section and go straight to the next section.

The classical zero-error capacity of a quantum channel, introduced in Chap. 5, was defined in terms of the clique number of the characteristic graph of a quantum channel. Now, in Sect. 8.2, we comment on the results introduced more recently: the literature by Scarpa, Severini, and Mancinska [26, 32]. Their contribution, mainly the second one, clearly binds Kochen-Specker (also known as Bell-Kochen-Specker) theorems to the quantum zero-error information theory.

A quantum version of the Wielandt's inequality [31] is described in Sect. 8.3. This inequality states an upper bound to the number of uses of a quantum channel in order to map an arbitrary density operator to a full rank operator. In this interesting paper, the authors state a remarkable relation with the quantum zero-error information theory intermediated by dichotomy theorems.

A variant of the zero-error capacity which considers entanglement assistance is presented in Sect. 8.4. Results of Winter et al. [14, 15, 17] in which non-commutative graphs are used for the quantum zero-error information theory are presented in Sect. 8.5. A quantum version of the Lovász theta function and some alternate definitions for the zero-error capacity of a quantum channel are presented as well.

A non-trivial application of zero-error quantum channels to help in determining the complexity class of a well-known problem was proposed by Beigi and Shor [5] which is now depicted in Sect. 8.5.

© Springer International Publishing Switzerland 2016
E.B. Guedes et al., *Quantum Zero-Error Information Theory*,
DOI 10.1007/978-3-319-42794-2_8

8.1 Bell's Inequalities

Entanglement motivated the famous article "Can Quantum Mechanical Description of Physical Reality Be Considered Complete?" written by Einstein, Poldosky, and Rosen, EPR for short [19]. After an important and long discussion, concepts like *principle of locality*, *elements of reality*, and *hidden variables* were defeated. The principle of locality, for example, claims that the events occurring in place are independent of parameters, eventually controlled at another "distant place" in the same time, but it was not confirmed.

The main assumption in EPR argument is the a priori concept of element of reality that could be obeyed by the Nature. The EPR paper aimed to show that quantum mechanics was an incomplete theory based on a *sufficient condition* for a physical property to be an element of reality:

> "If, without in any way disturbing a system, we can predict with certainty (i.e., with probability equal to unit) the value of a physical quantity, then there exists an element of physical reality corresponding to this physical quantity" [19].

Example 8.1 (Quantum Correlations are Stronger than the Classical Ones [29]). This example sets forth that quantum correlations are in general stronger than the classical ones. Consider a block of explosive material, in rest at $t = 0$, so at this time with angular moment $\vec{J} = 0$ exploding in two asymmetric parts, as shown in Fig. 8.1. Due to conservation laws [39, p. 323], the two parts carry angular moments, $\vec{J_1} = -\vec{J_2}$, respectively.

Suppose observers detecting the fragments and measuring the *classical dynamical variables* $a = \text{sign}(\vec{\alpha} \cdot \vec{J_1})$ and $b = \text{sign}(\vec{\beta} \cdot \vec{J_2})$, respectively, where $|\alpha\rangle$ and $|\beta\rangle$ are arbitrary unit vectors chosen by the observers. Obviously, $a, b = \pm 1$.

For N repetitions of the experiment, with directions of $\vec{J_1}$ and $\vec{J_2}$ randomly distributed, the averages are near to zero, that is,

$$\langle a \rangle = \frac{1}{N} \sum_{j=1}^{N} a_j \approx 0, \quad \langle b \rangle = \frac{1}{N} \sum_{j=1}^{N} b_j \approx 0. \tag{8.1}$$

Fig. 8.1 Classical setup with zero angular moment

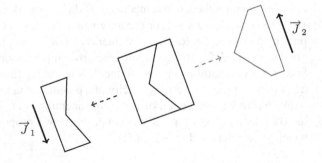

In order to compare their results, the observers calculate the *correlation*, defined by

$$\langle ab \rangle = \frac{1}{N} \sum_{j=1}^{N} a_j b_j. \tag{8.2}$$

The correlation is not zero in general. For concreteness, taking $\vec{\alpha} = \vec{\beta}$, the observers get $a_j = -b_j$, and in this case correlation yields $\langle ab \rangle = -1$.

For arbitrary $\vec{\alpha}$ and $\vec{\beta}$ the solution is [29]:

$$\langle ab \rangle = -1 + \frac{2\theta}{\pi}, \qquad 0 \le \theta \le \pi, \tag{8.3}$$

where θ stands for the angle between directions $\vec{\alpha}$ and $\vec{\beta}$. Notice that the correlation increases linearly from -1 to $+1$ as θ varies from 0 to π. Such correlation is shown in the plot of Fig. 8.2

Now let's consider the quantum turn. Consider the quantum analogy taking into consideration the singlet (entangled state)

$$|\psi\rangle = \frac{|01\rangle - |10\rangle}{\sqrt{2}}. \tag{8.4}$$

Assume that observers measure the observables $\vec{\alpha} \cdot \sigma_a$ along the axis $\vec{\alpha}$ and $\vec{\beta} \cdot \sigma_b$, along the axis $\vec{\beta}$, respectively. As before, in the classical analog, unit vectors are arbitrarily chosen by the observers and the possible values of a and b measurements are ± 1. The average values are both zero, that is

$$\langle a \rangle = \langle b \rangle = 0. \tag{8.5}$$

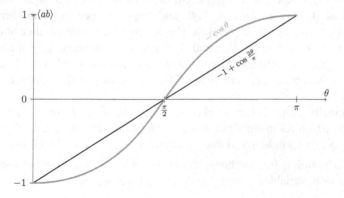

Fig. 8.2 Classical and quantum correlations

Furthermore, correlation can be calculated, according to quantum mechanics rules, as

$$\langle ab \rangle = \langle \psi | (\vec{\alpha} \cdot \sigma_a)(\vec{\beta} \cdot \sigma_b) | \psi \rangle , \tag{8.6}$$

where σ_a and σ_b are the Pauli matrices of the systems a and b, respectively.

For the singlet one has

$$\sigma_a | \psi \rangle = -\sigma_b | \psi \rangle . \tag{8.7}$$

Therefore, using the identity $(\vec{\alpha} \cdot \sigma)(\vec{\beta} \cdot \sigma) = \vec{\alpha} \cdot \vec{\beta} + \iota (\vec{\alpha} \times \vec{\beta}) \cdot \sigma$, the correlation is then obtained

$$\langle ab \rangle = - \cos \theta . \tag{8.8}$$

The main remark here is that quantum correlation, which is also shown in Fig. 8.2, is stronger than the classical correlation for all values of θ, except $\theta = 0$, $\frac{\pi}{2}$ and π.

The EPR paradox was solved by the *Bell's inequality*. It is interesting to remark that the inequality is not about quantum mechanics, rather its proof is general and independent of Physics. The central statement is that if one assumes validity of principle of locality, then there is an upper bound to the correlation between distant events. What Bell's inequality states is that local realism is incompatible with quantum mechanics.

In order to see an explanation why this happens, one must consider the thought experiment outlined in Fig. 8.3. Assuming the EPR principle, that is assuming the truth of local realism or the existence of hidden variables, we shall perform calculations to obtain the Bell's CHSH inequalities[1] [11]. After, real measurements demonstrate the violation of those inequalities.

In the thought experiment, a physicist, say, Charlie, repeats a large number, N, of preparations of two particles (say "*left*" and "*right*," respectively) and send, one by one, to his colleagues, Alice and Bob. The left particle is sent to Alice and the right one is sent to Bob. Later, Alice and Bob perform simultaneously measurements on their respective particles. Alice's lab is too far from Bob's lab in such a way their respective actions are *concurrent* [30], that is, their actions are *relativistically disconnected*.

Additionally, Alice is free to choose directions $\vec{\alpha}$ or $\vec{\beta}$ to perform her measurements, which results in a random variable denoted by $A = \pm 1$, if she chooses direction $\vec{\alpha}$, or a random variable denoted by $B = \pm 1$ if she chooses direction $\vec{\beta}$. Similarly, Bob is free to choose directions $\vec{\gamma}$ or $\vec{\delta}$, and from his measurement obtains random variables $C = \pm 1$ and $D = \pm 1$, respectively.

---

[1] The letters CHSH are a mention to the authors of this form of Bell's inequality [11].

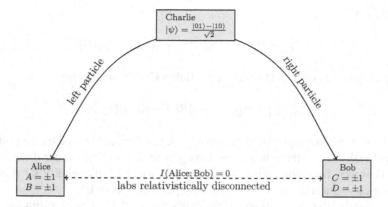

Fig. 8.3 Bell's inequality

Table 8.1 Values assumed by the random variables A and B

	$A = -1$	$A = 1$		$A = -1$	$A = 1$
(a) Values for $A + B$			(b) Values for $A - B$		
$B = -1$	-2	0	$B = -1$	0	2
$B = 1$	0	2	$B = 1$	-2	0

Now consider the random variable V defined by the following sum

$$V = AC + BC + BD - AD \qquad (8.9)$$

$$= (A + B)C + (B - A)D, \qquad (8.10)$$

where (8.10) follows from a simple rearrangement. From (8.10), it is clear that, as $A = \pm1$ and $B = \pm1$, either $(A + B)C = 0$ or $(B - A)B = 0$. From this, according to Table 8.1, it is easy to check that

$$V = AC + BC + BD - AD = \pm2. \qquad (8.11)$$

Now consider the expected value $\mathbb{E}[V]$:

$$\mathbb{E}[V] = \mathbb{E}[AC + BC + BD - AD] \qquad (8.12)$$

$$= \sum_{a,b,c,d} \Pr[a, b, c, d](ac + bc + bd - ad) \qquad (8.13)$$

$$\leq 2 \sum_{a,b,c,d} \Pr[a, b, c, d] \qquad (8.14)$$

$$\leq 2, \qquad (8.15)$$

where (8.12) and (8.13) are definitions, (8.14) is justified by (8.11), and (8.15) is because probabilities sum to one. On the other hand, from linearity of the expectation

$$\mathbb{E}[V] = \mathbb{E}[AC] + \mathbb{E}[BC] + \mathbb{E}[BD] - \mathbb{E}[AD] \qquad (8.16)$$

Comparing (8.15) with (8.16), we get the Bell's CHSH inequality:

$$\mathbb{E}[AC] + \mathbb{E}[BC] + \mathbb{E}[BD] - \mathbb{E}[AD] \le 2. \qquad (8.17)$$

Recall that the last inequality, shown in (8.17), was obtained based on the principle of local realism, and there is nothing wrong with this formula under that assumption. However, here, the authors objective was to check its validity for quantum mechanics. A question that arises is: how to perform this task?

Fortunately, the expectations on the left side of (8.17) can be estimated, with accuracy $\frac{1}{\sqrt{N}}$, through repeating the experiments N times. For example, let $\widehat{\mathbb{E}[AC]}$ denote the estimate of $\mathbb{E}[AC]$, then

$$\widehat{\mathbb{E}[AC]} = \frac{\sum_{j=1}^{N} a_j c_j}{N} \longrightarrow \mathbb{E}[AC], \qquad (8.18)$$

with high probability, as the number of repetitions, N, increases [28].

Now, consider quantum mechanics into account. As it is suggested in Fig. 8.3, Charlie sends the left qubit to Alice and the right qubit to Bob. The random variables A, B, C, and D are defined by the results of measurements of the following observables:

$$A = Z_1 \qquad C = \frac{-Z_2 - X_2}{\sqrt{2}} \qquad (8.19)$$

$$B = X_1 \qquad D = \frac{Z_2 - X_2}{\sqrt{2}} \qquad (8.20)$$

where subscripts 1, 2 stand for the left and right qubits sent by Charlie, respectively. The calculation of the expected values is straightforward, for example,

$$\mathbb{E}[AC] = \langle \psi | \left(Z_1 \otimes \frac{-Z_2 - X_2}{\sqrt{2}} \right) | \psi \rangle , \qquad (8.21)$$

and similar to BC, BD, and AD. But these calculations turn out to:

$$\mathbb{E}[AC] = \mathbb{E}[BC] = \mathbb{E}[BD] = -\mathbb{E}[AD] = \frac{1}{\sqrt{2}}. \qquad (8.22)$$

But, with these values, the sum adds up to:

$$\mathbb{E}[AC] + \mathbb{E}[BC] + \mathbb{E}[BD] - \mathbb{E}[AD] = 2\sqrt{2}. \qquad (8.23)$$

The result obtained in (8.23) shows a clear violation to the upper bound obtained in (8.17).

The conflicting results of Bell's CSHS inequality and the last result obtained from quantum mechanics vide (8.23) only can be solved via experimental procedures. Several such procedures were performed starting in the decades of 1960 and 1970. One of the most important was the work of Aspect et al. [3] that used two-photon atomic transitions in the setup. The results corroborated the predictions of quantum mechanics.

What the prior description shows is that for entangled states it is viable to find a pair of observables correlated in such a way their correlations violate the Bell's inequality. The meaning is that quantum mechanics produces statistical predictions that cannot be explained if one assumes the *Einstein locality*, that is, assuming that the results of experiments performed in a location are independent of another, discretionary one performed in another distant location, simultaneously.

8.1.1 Functional Consistency

Due to the complexity in demonstrating mentioned violations of Bell's CHSH inequality, new ways for demonstrated nonlocality were proposed [27]. Given a set of commuting observables $A, B, C, \ldots$ and a set of quantum states $|\psi\rangle, |\phi\rangle, \ldots$, then it is viable measuring the observables simultaneously and to obtain the joint distribution of the values of the observables chosen from that set. Consider an ensemble of identically prepared systems, in the state, say, $|\phi\rangle$, and suppose these states are described by observables $A, B, C, \ldots$. Each measure shall assign numerical values for each observable, $v(A), v(B), v(C), \ldots$. Quantum rules require that $v(A)$ is dependent only on the operator A, not on the state $|\phi\rangle$, and require also that in a commuting set of observables the only allowed results of simultaneous measurement are in the set of simultaneous eigenvalues.

From requirements considered [27], it is possible to notice that for any particular functional identity

$$f(A, B, C, \ldots) = 0, \tag{8.24}$$

fulfilled by the commuting observables, should be fulfilled by the set of eigenvalues, that is

$$f(v(A), v(B), v(C), \ldots) = 0. \tag{8.25}$$

For example, if A and B commute, then

$$C = A + B \stackrel{([A,B]=0)}{\Longrightarrow} v(C) = v(A) + v(B), \tag{8.26}$$

or, equivalently,

$$C - A - B = 0 \overset{([A,B]=0)}{\implies} v(C) - v(A) - v(B) = 0. \tag{8.27}$$

But, implications (8.26) or (8.27) are valid only if A and B commute! Because if it is not this way, A's and B's eigenvalues are different and they cannot be simultaneously measured. There is no evidence supporting those identities. However, in the sense of mean, (8.27) holds ever, that is, for any quantum state $|\phi\rangle$, it is true that

$$\langle \phi | A + B | \phi \rangle = \langle \phi | A | \phi \rangle + \langle \phi | B | \phi \rangle, \tag{8.28}$$

even despite the fact that A and B do not commute. Historically, famous mistakes happened probably motivated by this caveat [36].

We have seen that one important meaning of the Bell's CHSH inequalities is that for an entangled quantum state it is viable to find pairs of observables such that quantum mechanics statistics predictions are incompatible with the requirement of *locality* (also referred to as *Einstein locality*). Equivalently, this means that the results of measurements made at a given place are not all independent of those obtained at a remote lab.

An $n \times n$ matrix M is said to be diagonalizable if and only if the sum of the dimensions of the its eigenspaces equals n or, equivalently, if and only if M is similar to a diagonal matrix, that is, there exist an invertible matrix P such that

$$P^{-1}MP = \begin{pmatrix} \lambda_1 & 0 & \dots & 0 \\ 0 & \lambda_2 & \dots & 0 \\ \vdots & \vdots & \vdots & \vdots \\ 0 & 0 & \dots & \lambda_n \end{pmatrix}, \tag{8.29}$$

where the scalars λ_i ($i = 1, 2, \dots n$) are the eigenvalues of M. From

$$MP = P \begin{pmatrix} \lambda_1 & 0 & \dots & 0 \\ 0 & \lambda_2 & \dots & 0 \\ \vdots & \vdots & \vdots & \vdots \\ 0 & 0 & \dots & \lambda_n \end{pmatrix} = (\lambda_1 |\alpha_1\rangle \ \lambda_2 |\alpha_2\rangle \ \dots \ \lambda_n |\alpha_n\rangle), \tag{8.30}$$

where P can be written as

$$P = (|v_1\rangle, |v_2\rangle, \dots, |v_n\rangle). \tag{8.31}$$

In this representation $|\alpha_i\rangle$ ($i = 1, 2, \dots, n$) and α_i ($i = 1, 2, \dots, n$) are the eigenvectors and eigenvalues of M, respectively.

If for a matrix M there is only one basis in which it is diagonal, that basis corresponds to a *maximal quantum test* which is equivalent to a measurement of the observable represented by M. If $[M, N] = 0$ (M and N commute), it is possible to find a basis in which both matrices are diagonal. This basis corresponds to a maximal test, which provides a measurement of both M and N. Therefore two commuting operators can be simultaneously measured, otherwise they are said to be *incompatible*.

Generalization is straightforward. A set of matrices are said to be simultaneously diagonalizable if there exists a single invertible matrix P such that $P^{-1}MP$ is a diagonal matrix for every M in the set. A set of diagonalizable matrices commutes if and only if the set is simultaneously diagonalizable. A set of commuting operators is said to be *complete* if there exists a single basis in which all these operators are diagonal.

8.1.2 Context

If, regardless of the previously mentioned ambiguity, we insist in assuming that the measurement of an operator M depends uniquely on the objective properties of the measured quantum system, then we are assuming validity of contextualization of the setup to determine the measurement results completely. For example, if $[M, N] = 0$ and also, $[M, V] = 0$, we can jointly measure M and N, or jointly M and V, then we wait that the result of the measurement of M does not depend on its *context*, specifically, whether we measure only M, or M and N, etc. Notice that the assumption made is clearly counterfactual, that is, it cannot be put under an experimental setup.

Considered jointly, *contextuality* and *functional consistency* are indeed incompatible with the predictions of quantum theory, in spite of their "reasonability." The following example is particularly illustrative of this point.

Example 8.2 (Peres [29]). Consider a pair of qubits, not necessarily entangled (singlets), and the operators displayed in the following matrix

$$\mathcal{A} = \begin{pmatrix} \mathbb{1} \otimes Z & Z \otimes \mathbb{1} & Z \otimes Z \\ X \otimes \mathbb{1} & \mathbb{1} \otimes X & X \otimes X \\ X \otimes Z & Z \otimes X & Y \otimes Y \end{pmatrix}. \tag{8.32}$$

It is possible to verify that:

- Each operator has eigenvalue ± 1;
- In each row the three operators commute;
- In each column the three operators commute;
- Each operator is product of the two others, with exception of the third column, that requires a minus sign.

On the other hand, consider the sequences

$$(X \otimes Z)(Z \otimes X) = (XZ) \otimes (ZX) \tag{8.33}$$

$$= (-\imath Y) \otimes (\imath Y) \tag{8.34}$$

$$= Y \otimes Y, \tag{8.35}$$

and

$$(Z \otimes Z)(X \otimes X) = (ZX) \otimes (XZ) \tag{8.36}$$

$$= (-\imath Y) \otimes (-\imath Y) \tag{8.37}$$

$$= -Y \otimes Y. \tag{8.38}$$

Due to the opposite signs in (8.35) and (8.38), we cannot assign ± 1 values to entries of matrix $\mathcal{A}$ in such a way that those came out of measuring the operators form $\mathcal{A}$.

8.1.3 Gleason's Theorem

This important theorem states that for Hilbert spaces of dimension at least three, the only possible probability measures are that of the form

$$\langle A \rangle = \mathrm{Tr}\,(\rho A), \tag{8.39}$$

where ρ stands for a prepared quantum state and A is an observable. This means that there is not observable other than (8.39). The proof of that theorem is recognized as difficult.[2]

One of the issues raised by Gleason is that assuming only the primitives:

- Decision tests (only yes/no answers allowed) are represented by projectors in a Hilbert space;
- Compatible tests simultaneously correspond to commuting projectors;
- If P and Q are orthogonal projectors, then their sum, $S = P + Q$, which is itself a projector, obeys

$$\langle S \rangle = \langle P \rangle + \langle Q \rangle, \tag{8.40}$$

then, for Hilbert spaces with dimension larger than 2, (8.39) is the only that gives the corrected statistics for the measurements. The main remark is that the projector S, $\mathrm{Tr}\,S = 2$, can be decomposed in unlimited number of manners. For instance, take

$$P_1 = |\alpha\rangle \langle \alpha|, \tag{8.41}$$

$$P_2 = |\beta\rangle \langle \beta|, \tag{8.42}$$

---

[2]See an interesting "geometry oriented" discussion in [6, p. 151].

projectors onto orthonormal vectors $|\alpha\rangle$ and $|\beta\rangle$, respectively. Consider the next "rotations":

$$|a\rangle = \frac{|\alpha\rangle + |\beta\rangle}{\sqrt{2}}, \tag{8.43}$$

$$|b\rangle = \frac{|\alpha\rangle - |\beta\rangle}{\sqrt{2}}. \tag{8.44}$$

Clearly $|a\rangle$ and $|b\rangle$ are also orthonormal. The projectors onto these last vectors are

$$Q_1 = |a\rangle\langle a|, \tag{8.45}$$

$$Q_2 = |b\rangle\langle b|. \tag{8.46}$$

The last two decompositions satisfy

$$Q_1 + Q_2 = |a\rangle\langle a| + |b\rangle\langle b| \tag{8.47}$$

$$= \frac{(|\alpha\rangle + |\beta\rangle)(\langle\alpha| + \langle\beta|)}{2} + \frac{(|\alpha\rangle - |\beta\rangle)(\langle\alpha| - \langle\beta|)}{2} \tag{8.48}$$

$$= P_1 + P_2. \tag{8.49}$$

The identity $Q_1 + Q_2 = P_1 + P_2$ is considered as trivial, but, in contrast, the similar statement about the averages is not. Such statement regarding the averages can be formally written as:

$$\langle P_1\rangle + \langle P_1\rangle = \langle Q_1\rangle + \langle Q_2\rangle. \tag{8.50}$$

Considering such non-trivial nature, it deserves to be experimentally verified [29].

8.1.4 The Kochen-Specker Theorem

Mermin [27] introduced the following reasoning. Take observables with eigenvalues 1 or 0 with corresponding spin components $0, \pm 1$. The sums of the squared spin components along any three orthogonal axis, x, y, z obey

$$S_x^2 + S_y^2 + S_z^2 = s(s+1) = 2. \tag{8.51}$$

This is valid for particles with spin $s = 1$. Additionally, the squared components S_x^2, S_y^2, and S_z^2 form a mutually commuting set. The results of measurement are 0 or 1 for each direction, x, y, or z, additionally, that results must fulfill condition (8.51).

Assume that a *set of directions* with many orthogonal trials is given, in conjunction with the set of observables (squared spins components) alongside that directions. As the observables along orthogonal axes, the squared components are mutually commuting and due to constraint (8.51), the measured values equals one of the three triads $(1, 1, 0)$, $(1, 0, 1)$ or, $(0, 1, 1)$.

The contextuality, that is, the impossibility of description supported on *local hidden variables*, can be proved revealing quantum states for which the statistics of their respective observables (S_x^2, S_y^2, S_y^2), connected with the orthogonal axes, cannot be obtained by any assignments of 1s and 0s to every direction in the set, such that condition (8.51) holds.

The Kochen-Specker theorem exhibits a set of vectors, called *Kochen-Specker (KS) sets* such that it is impossible to assign 1's (associated with a color red) and 0's (associated with a color blue) and condition (8.51) is kept.

Notice that no statistics relative of the states is necessary for justification. This exclusion of statistics from the problem is similar to the change from the classical information theory, where asymptotically small probability of error is admitted to the zero-error information theory where no error is admitted. In the first, the ordinary information theory, probability measures are essential. For the last, zero-error information theory, graph theory and combinatorics are the main tools for analysis.

From the last discussion, it is possible to give a concise statement of the KS theorem in terms of the following problem.

Problem 8.1 (Mermin [27]). Determine a set of directions (vectors) in a 3-dimensional space such that it is impossible to assign a color (red or blue) to each direction under the condition that every subset of three mutually orthogonal vectors contains exactly one blue and two red vectors.

For the sake of completeness, the solution (proof) given by Mermin [27] is sketched here. Firstly, notice that only directions are essential, one is free to modify the size of the vectors. Without loss of generality assume the unit vector $\mathbf{z}$, blue, defining this axis. Take the red vector $\mathbf{a}$ living in the y-z plane:

$$\mathbf{a} = \mathbf{z} + \alpha\mathbf{y}, \ 0 < \alpha < 0.5. \tag{8.52}$$

Then consider the following remarks:

- As $\mathbf{z}$ is blue, $\mathbf{x}$ and $\mathbf{y}$ are both red. Indeed, any vector in the x-y must be red, due to the condition that one cannot have two orthogonal blue vectors, that is

$$\mathbf{c} = \beta\mathbf{x} + \mathbf{y}, \text{ for any } \beta. \tag{8.53}$$

  must be red.
- Additionally, since $\mathbf{a}$ and $\mathbf{x}$ are red, any vector in their plane must be red. To the proof, we shall soon verify that an interesting red vector in this plane is

$$\mathbf{d} = \frac{1}{\beta}\mathbf{x} - \frac{1}{\alpha}\mathbf{a}. \tag{8.54}$$

- Notice that because $\mathbf{a} = \mathbf{z} + \alpha\mathbf{y}$, then d is orthogonal to $\mathbf{c} = \beta\mathbf{x} + \mathbf{y}$. To see this, it is enough to perform the scalar product (denoted by "$\cdot$"):

$$\mathbf{d} \cdot \mathbf{c} = \left(\frac{1}{\beta}\mathbf{x} - \frac{1}{\alpha}\mathbf{a} \right) \cdot (\beta\mathbf{x} + \mathbf{y}) \tag{8.55}$$

$$= \|\mathbf{x}\|^2 - \frac{\beta}{\alpha}(\mathbf{z} + \alpha\mathbf{y}) \cdot \mathbf{x} - \frac{1}{\alpha}(\mathbf{z} + \alpha\mathbf{y}) \cdot \mathbf{y} \tag{8.56}$$

$$= \|\mathbf{x}\|^2 - \|\mathbf{y}\|^2 \tag{8.57}$$

$$= 0. \tag{8.58}$$

- Recall that $\mathbf{c}$ and $\mathbf{d}$ are both red, so the normal to their plane must be blue. Therefore, any vector in their plane must be red. So, the following sum is red:

$$\mathbf{e} = \mathbf{d} + \mathbf{c}. \tag{8.59}$$

- Since $\alpha \in (0, 0.5)$, then $\frac{1}{\alpha} > 2$, and, for $\beta \in \mathbb{R}$,

$$\left| \beta + \frac{1}{\beta} \right| \in (2, \infty), \tag{8.60}$$

it is viable determined a value of β such that $\mathbf{e}$ is parallel to

$$\mathbf{f} = \mathbf{x} - \mathbf{z}, \tag{8.61}$$

Also, changing the signal of β results in another $\mathbf{e}$ parallel to

$$\mathbf{g} = -\mathbf{x} - \mathbf{z}. \tag{8.62}$$

- Since $\mathbf{e}$ is red independently of the β value, both $\mathbf{f}$ and $\mathbf{g}$ must be red.
- But, $\mathbf{f} \cdot \mathbf{g} = 0$, they are orthogonal; so the normal to their plane is blue and any vector in their plane is surely red.
- Notice that

$$\mathbf{z} = -\frac{1}{2}\mathbf{f} - \frac{1}{2}\mathbf{g} \tag{8.63}$$

lives in the $\mathbf{f}$, $\mathbf{g}$ plane, but $\mathbf{z}$ is blue. This is the contradiction sought.

A chain of simpler and simpler proofs of quantum contextuality has been introduced since the KS theorem has appeared in the literature. One of these recent simplifications was introduced by Cabello et al. [9]. Experimental apparatus has also been explored in this same kind of sequence of simplifications [16].

An interesting approach concerning the plethora of nonlocality proofs is given in van Dam et al. [35], where the authors, rooted on Kullback-Lieber distance, propose a measure for the amount of evidence provided by the experimental setup.

In the following sections, recent results related with quantum zero-error information theory are examined.

8.2   Quantum Chromatic Number and Kochen-Specker Sets

Throughout this section we shall use definitions from graph theory introduced earlier in Chap. 4. Eventually, for the sake of easiness, some of those may be redundant.

Scarpa and Severini [32] and Mancinska et al. [26] introduced conditions for equality and strict inequality between three quantities associated with a graph G:

- The minimum dimension of orthogonal representation, denoted by $\xi(G)$;
- The quantum chromatic number, denoted by $\chi_q(G)$;
- The *ordinary* quantum chromatic number, denoted by $\chi(G)$.

One remarkable contribution introduced in the mentioned works is the outstanding role performed by the KS sets in the proofs. These sets are collections of vectors with applications to investigations about the calculation of quantum zero-error capacities of quantum channels.

The *quantum chromatic number* is a remarkable parameter for at least one reason: it is a tool for differentiating aspects of quantum and classical behavior, in particular, for entanglement-assisted communications. Also the quantum chromatic eases comprehension of combinatorial parameters as, e.g., the Lovász theta function and the minimum dimension of an orthogonal representation of a simple graph.

In this section only simple graphs (unweighted, unidirected graphs without self-loops) are considered and, as before, for a graph G, $V(G)$ and $E(G)$ denote its vertex and edge sets, respectively.

Before introducing the relationship between the concepts of quantum chromatic number, KS sets, and their consequences for the zero-error information theory, however, due to its importance, we shall review main concepts related to the Kochen-Specker theorem, following mainly the reading given in Peres [29].

A proper k-coloring of a graph is an assignment of k colors such that every two adjacent vertices have different colors. The chromatic number of a graph G, $\chi(G)$, is the minimum number of colors k needed to build a proper $k-$coloring map of G.

We now introduce a coloring game for a graph $G = (V, E)$. We consider that Alice and Bob claim that they have a proper k-coloring for G and a referee tests this claim with a one-round game. The rules of the game do not allow communication between the players. The referee asks Alice the color, say a, for the vertex v and Bob for the color, say, b, for the vertex w. Alice and Bob win the game if, for $a, b \in \{1, \ldots, k\}$:

- IF $v = w$, THEN $a = b$;
- IF $(v, w) \in E$, THEN, $a \neq b$.

A classical strategy is formed by two deterministic functions:

$$g_A : V \rightarrow \{1, \ldots, k\}, \tag{8.64}$$

$$g_B : V \rightarrow \{1, \ldots, k\}. \tag{8.65}$$

It is clear that, independently of the strategy chosen by the players, including even probabilistic strategies, they cannot win the game with probability 1 if $k < \chi(G)$, that is, using less than the chromatic number of colors in their assumed proper k-coloring procedure.

A quantum strategy the players can take advantage of is a convenient entangled state $|\psi\rangle$ living in a Hilbert space of dimension d and two collections of POVMs in the following way:

- For all $v \in V$, Alice owns $\{\mathbf{E}_{va}\}_{a=1,\ldots,k}$ and similarly Bob owns $\{\mathbf{F}_{vb}\}_{b=1,\ldots,k}$;
- Alice applies her POVM $\{\mathbf{E}_{va}\}_{a=1,\ldots,k}$ to her part of the entangled state and get the value a;
- Bob applies his POVM $\{\mathbf{F}_{va}\}_{a=1,\ldots,k}$ to his part of the entangled state and get the value b.

In order to have consistent conditions, for a quantum strategy, the rules are adapted in such a way that Alice and Bob win the game, if only if

$$\forall v \in V, \ \forall a \neq b, \ \langle \psi | \mathbf{E}_{va} \otimes \mathbf{F}_{vb} | \psi \rangle = 0, \tag{8.66}$$

$$\forall (v, w) \in E, \ \forall a, \ \langle \psi | \mathbf{E}_{va} \otimes \mathbf{F}_{wa} | \psi \rangle = 0. \tag{8.67}$$

If these conditions are attained, the strategy is said to be a *winning strategy*. Notice that only the number of measurement operator is fixed, neither the dimension of the entangled stated nor the rank of the measurement operator is taken into consideration. This motivates the following definition.

Definition 8.1 (Quantum Chromatic Number [32]). For all graphs G, the quantum chromatic number $\chi_q(G)$ is the minimum number k such that there exists a quantum k-coloring of G.

In the following $\overline{W}$ stands for the complex conjugate of the complex matrix W, that is, the (i, j) entry of $\overline{W}$ is obtained from the corresponding (i, j) entry of W taking its complex conjugate. The *Hilbert-Schmidt product* of two complex matrices W and V is given, as usual, by

$$\langle W | V \rangle = \operatorname{Tr} WV^{\dagger}. \tag{8.68}$$

The normal form of a k-coloring emphasizes the simplicity of its structure.

Proposition 8.1 (Normality [32]). *If G has a quantum k-coloring, then there exists a quantum k-coloring of G in* normal form, *with the following properties:*

1. *All POVMs are projective measurements with k projectors of rank r;*
2. *The state $|\psi\rangle$ is the maximally entangled of local dimension rk;*
3. *For all pairs v, a, the projectors of Alice and Bob are conjugate, that is, $\mathbf{E}_{v,a} = \bar{\mathbf{F}}_{v,a}$*
4. *The consistency conditions can be represented as*

$$\forall \ (v, w) \in E, \ \ \forall \ a \in \{1, \dots k\}, \ \ \langle \mathbf{E}_{va}, \mathbf{E}_{wa} \rangle = 0. \tag{8.69}$$

The proof for this proposition was introduced by Scarpa and Severini [32]. The authors emphasize that the quantum chromatic number depends on the rank of the POVM elements adopted by Alice and Bob. This remark motivates the following definition.

Definition 8.2 (Rank-r Quantum Chromatic Number [32]). The rank-r quantum chromatic number $\chi_q^{(r)}(G)$ of G is the minimum number of colors k such that G has quantum k-coloring formed by projectors of rank r and a maximally entangled state of local dimension rk.

It can be observed that $\chi_q^{(r)}(G) \leq \chi_q^{(s)}(G)$ if $r \geq s$ [32]. For rank-1 quantum coloring, the dimension of the maximally entangled state equals k and, the rank-1 projectors for each vertice v, can be represented as the outer product

$$|e_{va}\rangle \langle e_{va}|, \ \ a \in \{1, \dots, k\}, \tag{8.70}$$

for an orthonormal basis $\{|e_{va}\rangle_{a \in \{1,\dots,k\}}\}$. Therefore (8.69) can be rewritten as:

$$\forall (v, w) \in E(G), \ \ \forall a \in \{1, \dots, k\}, \ \ \langle e_{va} | e_{wa} \rangle = 0. \tag{8.71}$$

If a quantum k-coloring of a graph $G(V, E)$ is given, then a *matrix representation* of G can be constructed with the map:

$$\Phi : V \to \mathbb{C}^{k \times k}, \tag{8.72}$$

such that for all $(v, w) \in E$ it is required

$$\text{diag}(\Phi(v)^\dagger \Phi(w)) = 0. \tag{8.73}$$

Here, diag(A) stands for the vector formed with the A diagonal entries. The map Φ is built taking for all $v \in V$ a unitary matrix U_v mapping the computational basis $\{|i\rangle_{i \in \{1,\dots,k\}}\}$ to $\{|e_{va}\rangle, a \in \{1, \dots, \}\}$. Notice that U_v is a $k \times k$ matrix and (8.71) condition assures that if (v, w) is an edge then the diagonal entries of $U_v^\dagger U_w$ are zero.

8.2.1 Relationship Between $\xi(G)$, $\chi_q^{(1)}(G)$ and $\chi(G)$

A k-dimensional *orthogonal representation* of $G = (V, E)$ is a map

$$\phi : V \to \mathbb{C}^k, \tag{8.74}$$

such that for all $(v, w) \in E$, the inner product $\langle \phi(v) | \phi(w) \rangle = 0$. The *orthogonal rank* of a graph G, denoted by $\xi(G)$, is defined as the minimum k such that an orthogonal representation of G in $\mathbb{C}^k$ exists.

For all graphs G the following inequalities hold [10]:

$$\xi(G) \le \chi_q^{(1)}(G) \le \chi(G). \tag{8.75}$$

If we have two graphs G and H, we can now define the *Cartesian product GH*.

- The vertex set $V(GH) = V(G) \times V(H)$ is the Cartesian product of the vertex sets of G and H;
- An edge between vertices (a_1, b_1), $(a_2, b_2) \in V(GH)$ is either

  - $a_1 = a_2$ and $(b_1, b_2) \in E(H)$, or
  - $(a_1, a_2) \in E(G)$ and $b_1 = b_2$.

Notice that a vertex in $V(GH)$ corresponds to a pair (a, b) of vertices where a is from G and b is from H.

Let K_k be a complete graph with k vertices. The next proposition clarifies the relation between the minimum dimension of orthogonal representation and the quantum chromatic number.

Proposition 8.2 (Scarpa and Severini [32]). *For all graphs G:*

$$\chi_q^{(1)} = \min\{k : \xi(GK_c) = k\}. \tag{8.76}$$

From this proposition follows a condition for equality between the rank-1 quantum chromatic number and orthogonal rank of a graph G.

Theorem 8.1 (Scarpa and Severini [32]). *For all graphs G:*

$$\chi_q^{(1)}(G) = \xi(G) \text{ if only if } \xi(GK_{\xi(G)}) = \chi(G). \tag{8.77}$$

8.3 Wielandt's Inequality

We have defined a quantum channel, denoted $\mathcal{E}$, as a *trace preserving completely positive linear map* (TPCP), that is,

$$\mathcal{E} : \mathcal{M}_{D \times D} \to \mathcal{M}_{D \times D}, \tag{8.78}$$

where $\mathcal{M}_{D\times D}$ is the space of the complex $D \times D$ matrices. The Kraus operators $W_k \in \{\mathcal{M}_{D\times D}\}_{k=1}^{d}$ are a versatile representation of quantum channel $\mathcal{E}_W$:

$$\mathcal{E}_W(\rho) = \sum_{k=1}^{d} W_k \rho W_k^{\dagger}. \tag{8.79}$$

Unitary operations, some kind of measurements, addition of uncorrelated quantum systems, substituting of a input state by other state are some operations in a quantum channel that can be well represented by Kraus operators.

Another useful representation, that allows modeling any operation by a unitary operation on a larger Hilbert space, is the Stinespring theorem:

Theorem 8.2 (Stinespring Theorem). *Let $\mathcal{E}$ be a trace-preserving quantum operation on a Hilbert space $\mathcal{H}$. Then there is an ancilla space $\mathcal{K}$ of dimension $\dim \mathcal{K} \leq (\dim \mathcal{H})^2$ so that for any fixed $|\chi\rangle \in \mathcal{K}$ there is a unitary transformation $\hat{U}$ on $\mathcal{H} \otimes \mathcal{K}$ with*

$$\mathcal{E}(\rho) = \mathrm{Tr}_{\mathcal{K}}\{\hat{U}(\rho \otimes |\chi\rangle\langle\chi|)\hat{U}^{\dagger}\}. \tag{8.80}$$

However, for quantum channels the representation by means of Kraus operators is actually more usual, at least for applications where discrete classical information are to be transmitted. In this section we discuss some relevant development of the notion of zero-error communications through quantum channels.

As usual we begin with a classical concept to later introduce extensions of that concept into the quantum framework. Recall the definition of a classical discrete memoryless channel (DMC) $(\mathcal{X} \times \mathcal{Y}, W(Y|X))$ for which $|\mathcal{X}| = |\mathcal{Y}| = D$. The main elements of a DMC are shown in Fig. 8.4. The matrix $W(Y|X)$ is defined as a stochastic matrix whose rows are indexed by the elements of $\mathcal{X}$ while the columns are indexed by those of the finite set $\mathcal{Y}$. The (x, y) entry of $W(Y|X)$ is the probability that $Y = y$ is received when $X = x$ is transmitted.

Source messages V are picked from a finite set (alphabet of the source) and are mapped by means of the encoder in codewords $X^n = (X_1, X_2, \ldots, X_n)$. Each X_i, $i = 1, 2, \ldots, n$, is transmitted through the memoryless channel that produces an output Y_i in such a way that

$$\Pr(Y^n = y^n | X^n = x^n) = \prod_{i=1}^{n} \Pr(Y_i = y_i | X_i = x_i). \tag{8.81}$$

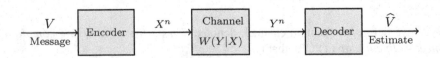

Fig. 8.4 Classical communication systems

The right-hand product means the i-th output y_i of a DMC depends by a stochastic map only on the i-th input x_i. This is the motivation to the term *memoryless* in the definition.

If we take into account the input and output blocks, X^n and Y^n, respectively, it is natural to define a stochastic matrix for these blocks, say,

$$W^{(n)}(Y^n|X^n) \triangleq \Pr(Y^n|X^n) = \underbrace{W \otimes W \otimes \ldots \otimes W}_{n \text{ times}}. \qquad (8.82)$$

Notice that the probability distributions $W(\cdot|x)$ and $W^{(n)}(\cdot|x^n)$ correspond to the x-th and x^n-th lines of the product matrix, respectively. If two input blocks X^n and $\tilde{X}^n$ can lead to the same output block Y^n, decoding cannot be performed without error. We say that X^n and $\tilde{X}^n$ are *confusable* or *indistinguishable*.

Sanz et al. [31] proposed an extension of the classical Wielandt's inequality to quantum channels. The concept concerns the number of applications of the channel to any source (probability distribution) for which any output will be reached.

Before introducing the formal definition for the quantum case, we will recall some important notions. A matrix is said to be *positive* if its entries are all strictly positive.[3] This means that for a positive DMC matrix W, any output $y \in \mathcal{Y}$ can be reached from any input $x \in \mathcal{X}$ at the input.

Definition 8.3 (Primitive Matrix). A square stochastic matrix W is said to be primitive if there is an $m \in \mathbb{N}$ such that $(W^m)_{ij} > 0$ for all (i,j), that is if W^m is positive. The minimum m for which this occurs, denoted $p(W)$, is said to be the classical index of primitivity of W.

It means that if a DMC probability transition matrix W is primitive, then using $p(W)$ times a DMC $(\mathcal{X} \times \mathcal{Y}, W(Y|X))$, any $y \in \mathcal{Y}$ can be reached from each input $x \in \mathcal{X}$ transmitted or, equivalently, for the product channel displayed in Fig. 8.4, all input blocks X^n are confusable.

The *Wielandt's inequality* [22, p. 520] states that, for every primitive matrix W, then:

$$p(W) \le D^2 - 2D + 2. \qquad (8.83)$$

Observe that the Wielandt's inequality does not depend on the matrix elements; only its primitivity is required. There are applications of the Wielandt's inequality for several fields, e.g., to graph theory, number theory, numerical analysis, etc. An extension of the concept of index of primitivity to the quantum framework was introduced by Sanz et al. [31] which is defined by the number of times a channel must be used, so that it maps any density operator to one with full rank.

---

[3]We call the attention that there is no connection of definition of *positive matrix* with the definition of *positive definite* matrix.

Fig. 8.5 Graph of the DMC induced by the stochastic matrix of (8.84) for $D = 5$. Labels on the edge (x, y) stand for conditional probabilities, for instance, $\alpha = \Pr[Y = 0|X = 4]$

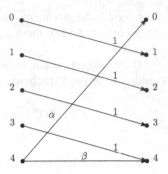

Example 8.3. The following $D \times D$ matrix

$$W = \begin{pmatrix} 0 & 1 & 0 & \cdot & \ldots & 0 & 0 \\ 0 & 0 & 1 & 0 & \ldots & & 0 \\ \vdots & \vdots & \vdots & \vdots & \vdots & \vdots & \vdots \\ 0 & 0 & 0 & 0 & \ldots & 0 & 1 \\ \alpha & \beta & 0 & 0 & \ldots & \cdot & 0 \end{pmatrix}$$

(8.84)

is primitive, for $\alpha > 0$ and $\beta > 0$, $\alpha + \beta = 1$. The primitivity is due to the fact that W^m is positive for $m = D^2 - 2D + 2$. This is just the Wielandt bound [28, p. 730]. For instance, fixing $D = 5$, we have $p(W) = m = 17$, which means that 17 uses of that DMC, every output block $Y^{17} \in \mathcal{Y}^{17}$ can be reached from any $X^{17} \in \mathcal{X}^{17}$ input block of symbols.

The graph of transitions for the DMC induced by this matrix $(D = 5)$ is displayed in Fig. 8.5.

Consider the probability row vector **p** of the input X as:

$$\mathbf{p} = (p_0, p_1, p_2, p_3, p_4).$$

(8.85)

That is, $p_x = \Pr[X = x]$, $x \in \mathcal{X}$. Similarly denote

$$\mathbf{q} = (q_0, q_1, q_2, q_3, q_4),$$

(8.86)

the probability row vector of the output Y, that is, $p_y = \Pr[Y = y]$, $y \in \mathcal{Y}$. The transition channel matrix W defines the relationship between probability vectors **p** and **q**, as follows:

$$\mathbf{q} = \mathbf{p}W.$$

(8.87)

8.3.1 Quantum Index of Primitivity

Let $\mathcal{E}_W$ be the quantum channel defined by Kraus operators $\{W_k \in \mathcal{M}_{D \times D}\}_{k=1}^d$, that is

$$\mathcal{E}_W(\rho) = \sum_{k=1}^d W_k \rho W_k^\dagger. \tag{8.88}$$

The quantum index of primitivity, denoted by q, is defined to a quantum channel (TPCP map) by the least $m \in \mathbb{N}$ such that m uses of the channel assures that every positive semidefinite operator is mapped onto a positive definite operator, where D is the dimension of the Hilbert space, and d the number of linearly independent Kraus operators.

It is possible to show that

$$q \leq (D^2 - d + 1)D^2, \tag{8.89}$$

where D is the dimension of the Hilbert space, and d the number of linearly independent Kraus operators.

Our proposal now is to focus on the issues of the inequality (8.89) connected with the notion of quantum channels with positive zero-error capacity. In this way, we need to recall some preliminaries given in [31]. Firstly, the authors define $S_n(W) \in \mathcal{M}_{D \times D}$ as the linear space spanned by all possible products of *exactly n* Kraus operators, $W_{k_1} W_{k_2} \ldots W_{k_n}$ and denote $W_k^{(n)}$ the elements of $S_n(W)$, with this, they define

$$H_n(W, \varphi) \triangleq S_n(W) |\varphi\rangle \subseteq \mathbb{C}^D, \tag{8.90}$$

as the space spanned by all vectors $W_{k_1} W_{k_2} \ldots W_{k_n} |\varphi\rangle$, where $|\varphi\rangle \in \mathbb{C}$. Secondly, Sanz et al. [31] recall the one-to-one correspondence between a quantum channel $\mathcal{E}$ and its *Choi matrix*

$$\omega(\mathcal{E}) \triangleq (\mathbb{1} \otimes \mathcal{E})(\Omega), \tag{8.91}$$

where $\Omega = \sum_{i,j}^D |ii\rangle \langle jj|$. Then, the observed rank $[\mathcal{E}_W^n(|\varphi\rangle \langle \varphi|)] = \dim [H_n(W, \varphi)]$. Equipped with prior discussion, three properties are introduced.

1. **Primitive Quantum Channel.** A quantum channel $\mathcal{E}_W$ is said to be *primitive* if there exists some $n \in \mathbb{N}$ such that for all $|\phi\rangle \in \mathbb{C}^D$, $H_n(W, \varphi) = \mathbb{C}^D$. The number $q(\mathcal{E}_W)$ stands for the minimum n for which the condition is reached. This means that for every input density operator ρ the output $\mathcal{E}_W^n(\rho)$, obtained after n applications of the channel is full-rank. It is observed that if $\mathcal{E}_W$ is primitive, then for every $m \in \mathbb{N}$, $\mathcal{E}_W^m$ is also primitive and we have

$$H_n(W, \varphi) = \mathbb{C}^D \text{ for all } n \geq q(\mathcal{E}_W). \tag{8.92}$$

2. **Eventually Full Kraus Rank Quantum Channel**. A quantum channel $\mathcal{E}_W$ is
   called *eventually full Kraus rank* if there exists some $n \in \mathbb{N}$ such that $S_n(W) = \mathcal{M}_{D \times D}$. This means that rank $\left[w(\mathcal{E}_W^n)\right] = D^2$. The number $i(W)$ stands for the
   minimum n for which that condition is satisfied. Notice that if $\mathcal{E}_W$ fulfills this
   property, then $S_n(W) = \mathcal{M}_{D \times D}$ for all $n \geq i(W)$.
3. **Strongly Irreducible Quantum Channel**. A quantum channel $\mathcal{E}_W$ is said to be
   strongly irreducible if the following two conditions are fulfilled:

   a. $\mathcal{E}_W$ has a unique eigenvalue, λ, with $|\lambda| = 1$;
   b. The corresponding eigenvector ρ is a positive definite operator ($\rho > 0$).

An important question now is how the classical Wielandt bound relates with the
quantum one. The main tool is to make an embedding of the classical channel in the
quantum framework, as it is shown in the next example.

Example 8.4. Consider again the classical DMC illustrated in Fig. 8.5 (for $D = 5$)
and respective stochastic matrix W given by (8.84). It is easy to see that embedding
is obtained by intermediate the following map $\mathcal{E}_W$ defined by the Kraus operators

$$W_{x,y} = \sqrt{w_{x,y}} \, |y\rangle \langle x| \, , \, x \in \mathcal{X}, y \in \mathcal{Y}. \tag{8.93}$$

For an input (diagonal) operator

$$\rho = \delta_{x,y} p_x = \begin{pmatrix} p_0 & 0 & 0 & 0 & 0 \\ 0 & p_1 & 0 & 0 & 0 \\ 0 & 0 & p_2 & 0 & 0 \\ 0 & 0 & 0 & p_3 & 0 \\ 0 & 0 & 0 & 0 & p_4 \end{pmatrix}. \tag{8.94}$$

Here, $\delta_{x,y}$ stands for the Kronecker function and p_x, $x \in \{0, 1, 2, 3, 4\}$, are the entries
of input probability vector $\mathbf{p}$ (recall Example 8.3). The output is given by

$$\sigma = \mathcal{E}_W(\rho) \delta_{x,y} q_y = \begin{pmatrix} q_0 & 0 & 0 & 0 & 0 \\ 0 & q_1 & 0 & 0 & 0 \\ 0 & 0 & q_2 & 0 & 0 \\ 0 & 0 & 0 & q_3 & 0 \\ 0 & 0 & 0 & 0 & q_4 \end{pmatrix}, \tag{8.95}$$

where q_y, $y \in \{0, 1, 2, 3, 4\}$, are the entries of the output vector $\mathbf{q}$.
   The set of Kraus operators are

$$\left\{ |1\rangle \langle 0| \, , \, |2\rangle \langle 1| \, , \, |3\rangle \langle 2| \, , \, |4\rangle \langle 3| \, , \, \sqrt{\alpha} \, |0\rangle \langle 4| \, , \, \sqrt{\beta} \, |4\rangle \langle 4| \right\}. \tag{8.96}$$

Assume W is a primitive DMC stochastic probability transition matrix with
primitivity index $p(W)$. The following is proved in [31].

Proposition 8.3. *Let W be a primitive stochastic map and let $\mathcal{E}_W$ be the corresponding TPCP. The channel $\mathcal{E}_W$ is also primitive and*

$$q(W) = p(W) = i(W). \tag{8.97}$$

Notice that equality of (8.97) holds for quantum channels resulting of the above DMC embedding. This fact is illustrated next.

Example 8.5. Let us consider a "genuine" quantum channel, with $D = 2$ and $d = 3$ given by the Kraus operators

$$\left\{ W_1 = \frac{1}{\sqrt{3}}X, \; W_2 = \frac{1}{\sqrt{3}}Y, \; W_3 = \frac{1}{\sqrt{3}}Z \right\} \tag{8.98}$$

where X, Y, and Z are the Pauli matrices. Explicitly, the map is given by

$$\mathcal{E}_W(\rho) = \frac{1}{3}\left(X\rho X^\dagger + Y\rho Y^\dagger + Z\rho Z^\dagger \right), \tag{8.99}$$

for an input state ρ.

It is straightforward to check that in this case $q(\mathcal{E}_W) = 1$ and $i(W) = 2$. The quantum Wielandt's bound is $8 = (D^2 - d + 1)D^2 \geq q(\mathcal{E}_W)$.

There is an open question if the quantum Wielandt's bound is sharp, however the following theorem is very important because establishing a *universal dichotomy* behavior of the zero-error capacity of the important class of unital quantum channels. This dichotomy result is universal in the sense that it depends only on the dimension of the Hilbert space, D, and not on the channel itself.

Theorem 8.3 (Dichotomy Behavior [31]). *Let $C^{(0)}(\mathcal{E})$ be the zero-error classical capacity of the quantum channel $\mathcal{E}$. If $\mathcal{E}$ is a quantum channel with full-rank fixed point, then either $C^{(0)}(\mathcal{E}^n) \geq 1$ for all n or $C^{(0)}\left(\mathcal{E}^{q(\mathcal{E})}\right) = 0$.*

Notice that if $\mathcal{E}^n$ stands for the input–output relationship after n units of time or space then the theorem reveals the existence of a *universal critical length* $n = q(\mathcal{E})$ such that once a transmission is successfully viable then a successful transmission $m \geq n$ is possible.

8.4 Entanglement-Assisted Zero-Error Capacity

In Sect. 3.2.5 we saw different capacities of quantum channels. The entanglement-assisted capacity in particular considers that the parties share an unrestricted amount of previously shared entanglement which they can use in order to maximize the information changed through the quantum channel. In this section, we describe how entanglement can be used in a zero-error scenario for exchanging classical information.

Before introducing the entanglement-assisted zero-error communication, we need some background concepts. We start with the hypergraph of a quantum channel.

Definition 8.4 (Hypergraph of a Quantum Channel [13]). Let $\mathcal{E}$ be a quantum channel. The hypergraph of $\mathcal{E}$, denoted by $\mathbf{H}(\mathcal{E})$, is a set of vertices, denoted by S and a set of subsets of S. The set of vertices S is composed by the channel input. There is one hyperedge for each of the outputs, which contains all the inputs that have a nonzero probability of causing that output.

We also use the concept of clique in hypergraphs. A clique of $\mathbf{H}(\mathcal{E})$ is a set κ_i of possible inputs of a given output in a confusability graph. In other words, the clique κ_i contains all the inputs that can cause the same output.

In this current scenario, in particular, prior to information exchange through the quantum channel, Alice and Bob share a d-dimensional entangled state $\rho_{AB} = |\Phi_{AB}\rangle \langle \Phi_{AB}|$ given by

$$|\Phi_{AB}\rangle = \frac{1}{\sqrt{d}} \sum_{i=0}^{d-1} |i_A\rangle \, |i_B\rangle \tag{8.100}$$

Considering that such pre-shared entanglement is available, Cubitt et al. [13] proposed a protocol for entanglement-assisted zero-error communication that is described as follows:

1. Alice chooses a message $m \in \{1, \ldots, K\}$ from a set of messages, where K is the number of messages. Alice wants to send the message m to Bob;
2. Alice measures her half of the entangled system using a complete orthogonal basis, say $B_m = \{|\psi_{\hat{x}}\rangle\}$, where $\hat{x}$ is a vertice in a clique κ_m from the hypergraph $\mathbf{H}(\mathcal{E})$;
3. Alice sends the result of her measurement to Bob.
   Some clarifications are needed before proceeding. In the hypergraph $\mathbf{H}(\mathcal{E})$ the vertice x represents the unit vector $|\psi_{\hat{x}}\rangle$ such that if x and $\hat{x}$ are adjacent then $\langle \psi_x| \psi_{\hat{x}}\rangle = 0$. Recalling that K is the size of the messages set, the hypergraph has K cliques of size d, say $\{\kappa_1, \ldots, \kappa_K\}$. It is analogous to say that each message m has a d-size clique κ_m in the hypergraph $\mathbf{H}(\mathcal{E})$.
4. After Alice's measurement, Bob's state will collapse to $|\psi_x\rangle^*$;
5. Bob will measure his state in $B_m = \{|\psi_x\rangle\}$ in order to get the final state $|\psi_{\hat{x}}\rangle^*$;
6. Bob output is denoted by y. His possible states are determined by those vertices x, for which $p(y|\hat{x}) > 0$ and these adjacent states are mutually orthogonal, i.e., for any $\hat{x}_1$ and $\hat{x}_2$, then $\langle \hat{x}_1| \hat{x}_2\rangle = 0$ [23].

For short, to send K messages using entanglement, Alice and Bob can use a maximally entangled state of rank d: to send m, Alice measures her side of the state in the bases B_m and obtains the outcome j (at random). She inputs $(m; j)$ to the channel. Bob's output tells him that Alice's input was in some particular mutually confusable subset, but by construction, these inputs correspond to mutually orthogonal residual states of his subsystem, so he can perform a projective measurement to determine precisely which input Alice made to the classical channel, and hence which of the K messages she chose to send, with certainty [13].

Using the previously defined elements and protocol characterized, we can now characterize the entanglement-assisted zero-error classical capacity of quantum channels.

Theorem 8.4 (Entanglement-Assisted Zero-Error Capacity [13]). *Let $\mathcal{E}$ be a quantum channel. The entanglement-assisted zero-error capacity of $\mathcal{E}$, denoted by $C_E^{(0)}(\mathcal{E})$, is given by*

$$C_E^{(0)}(\mathcal{E}) = \lim_{n\to\infty} \frac{1}{n} \log K^E(\mathcal{E}^{\otimes n}) \geq \log(K^E) \tag{8.101}$$

where K^E is the number of mutually non-adjacent input messages with entanglement assistance.

Theorem 8.5 (Cubitt-Leung-Matthews-Winter Theorem [13] via [23]). *For a quantum channel $\mathcal{E}$ with hypergraph $\mathbf{H}(\mathcal{E})$, there exists an entanglement-assisted quantum communication protocol that can send one of K messages with zero error; hence for entanglement-assisted asymptotic classical zero error capacity*

$$\log(K) \leq C^{(0)}(\mathcal{E}) = \lim_{n\to\infty} \frac{1}{n} \log(K(\mathcal{E}^{\otimes n})) < C_E^{(0)}(\mathcal{E})$$

$$= \lim_{n\to\infty} \frac{1}{n} \log K^E(\mathcal{E}^{\otimes n}) \geq \log(K^E). \tag{8.102}$$

This theorem shows us that entanglement can sometimes be used to increase the number of classical messages which can be sent perfectly over quantum channels [13].

Some results in the literature have interesting connections with the entanglement-assisted zero-error capacity. Leung et al. [25], using certain input codewords (based on a Pauli graph), show that entanglement can help to increase the classical zero-error capacity to the maximum achievable HSW capacity.

In general, it is possible to observe the following relation between the classical zero-error quantities:

$$C^{(0)}(\mathcal{E}) < C_E^{(0)}(\mathcal{E}) \leq C_{1,\infty}(\mathcal{E}) \tag{8.103}$$

Recalling that the zero-error capacity is a quantity hard to compute even for small characteristic graphs, upper bounds play an important role, in particular, the Lovász ϑ function is commonly considered. Beigi [4] verified that the Lovász ϑ function is an upper bound on the zero-error capacity even in the presence of entanglement between the sender and receiver.

8.5 Non-Commutative Graphs and Quantum Lovász ϑ Function

The works of Duan, Cubitt, Severini, Winter, and other collaborators [14, 15, 17] introduce the theory of non-commutative graphs in the study of quantum zero-error capacity problem. Starting with the Kraus form of representation of a quantum channel, the authors define a generalization of the classical adjacency graph called *non-commutative (confusability) graph* by the operator space:

$$S \triangleq \text{span}\{W_j^\dagger W_k : j, k\} < \mathcal{L}(\mathcal{H}) \tag{8.104}$$

where $\mathcal{L}(\mathcal{H})$ stands for the set of observable on Hilbert space $\mathcal{H}$.

According to this graph definition, a zero-error code consists in the anti-clique of the corresponding graph. The biggest anti-clique, called *independence number* and denoted by $\alpha(G)$, corresponds to the maximum number of messages that can be transmitted through the channel with probability of error equal to zero. This way, the classical zero-error capacity of a channel with graph G is given by

$$C^{(0)}(G) = \lim_{n \to \infty} \log \alpha(G^n) = \sup_n \log \alpha(G^n). \tag{8.105}$$

Translating this concept to the quantum scenario, we have a quantum channel $\mathcal{E} : \mathcal{B}(\mathcal{H}_X) \to \mathcal{B}(\mathcal{H}_Y)$, where $\mathcal{B}(\cdot)$ is the space of linear operators from a given Hilbert space. This way, the event $E_{x,y} : \mathcal{H}_X \to \mathcal{H}_Y$, that corresponds to the input of a quantum state $x \in \mathcal{X}$ and the output of a quantum state $y \in \mathcal{Y}$ in this quantum channel, is given by

$$E_{x,y} = \sqrt{p(y|x)} \, |y\rangle \, \langle x| \,. \tag{8.106}$$

This way, we can define the *confusability graph of a quantum channel* (or non-commutative graph) as being a subspace

$$S = \text{span} \left\{ E_{x',y'}^\dagger \cdot E_{x,y} \neq 0; x, x' \in \mathcal{X}, y, y' \in \mathcal{Y} \right\}. \tag{8.107}$$

It is interesting to notice that such definition emphasizes the channel's input that can be confused, while in Chap. 5 the approach was to emphasize inputs that are not adjacent. Despite this difference, both definitions are equivalent.

Example 8.6 (Confusability Graph of a Quantum Channel). Let $\mathcal{E}$ be a quantum channel shown in Fig. 8.6a. The input alphabet contains the symbols $\mathcal{X} = \{a, b\}$ and the output alphabet contains the symbols $\mathcal{Y} = \{c, d\}$.

According to (8.106), we have the following events:

$$E_{a,c} = 1 \cdot |c\rangle \, \langle a| = |c\rangle \, \langle a| \,, \tag{8.108}$$

Fig. 8.6 Example of a graph of a quantum channel (**a**) transitions probabilities at the channel's end (**b**) confusability graph

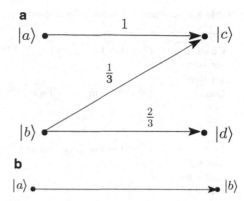

$$E_{a,d} = 0 \cdot |d\rangle \langle a| = 0, \tag{8.109}$$

$$E_{b,c} = \sqrt{\frac{1}{3}} |c\rangle \langle b|, \tag{8.110}$$

$$E_{b,d} = \sqrt{\frac{2}{3}} |d\rangle \langle b|. \tag{8.111}$$

It is important to emphasize that these events have a straight correspondence with the quantum channel it is related. From these events, we can consider the following elements that will compose the confusability graph (see (8.107))

$$E_{a,c}^{\dagger} \cdot E_{a,c} = |a\rangle \langle a|, \tag{8.112}$$

$$E_{a,c}^{\dagger} \cdot E_{b,c} = \sqrt{\frac{1}{3}} |a\rangle \langle b|, \tag{8.113}$$

$$E_{b,c}^{\dagger} \cdot E_{a,c} = \sqrt{\frac{1}{3}} |b\rangle \langle a|, \tag{8.114}$$

$$E_{b,c}^{\dagger} \cdot E_{b,c} = \frac{1}{3} |b\rangle \langle b|, \tag{8.115}$$

$$E_{b,d}^{\dagger} \cdot E_{b,d} = \frac{2}{3} |b\rangle \langle b|. \tag{8.116}$$

Thus,

$$S = \mathrm{span} \left\{ E_{a,c}^{\dagger} \cdot E_{a,c}, E_{a,c}^{\dagger} \cdot E_{b,c}, E_{b,c}^{\dagger} \cdot E_{a,c}, E_{b,c}^{\dagger} \cdot E_{b,c}, E_{b,d}^{\dagger} \cdot E_{b,d} \right\}. \tag{8.117}$$

Considering the subspace S we can denote the zero-error capacity of a quantum channel. This capacity is given by the following expression, considering that the biggest set of self-orthogonal states is given by $\{|\phi_m\rangle : m = 1, \dots, N\}$

$$\forall m \neq m' : |\phi_m\rangle \langle \phi_{m'}| \in S^{\perp}, \tag{8.118}$$

Table 8.2 Alternative definitions for zero-error capacities considering independence numbers

Capacity	Expression	Observations
Classical zero-error capacity	$C^{(0)}(S) = \lim_{n\to\infty} \frac{1}{n} \log \alpha(S^{\otimes n})$	Shown previously in (8.105)
Quantum zero-error capacity	$Q^{(0)}(S) = \lim_{n\to\infty} \frac{1}{n} \log \alpha_q(S^{\otimes n})$	α_q denotes the quantum independence number whose value depends on the existence of a Stienespring dilatation in the channel.
Entanglement assisted zero-error capacity	$C_E^{(0)}(S) = \lim_{n\to\infty} \frac{1}{n} \log \tilde{\alpha}(S^{\otimes n})$	$\tilde{\alpha}$ denotes the higher integer N for which there are (i) Hilbert spaces $\mathcal{H}_{X0}$ and $\mathcal{H}_{Y0}$; (ii) $\omega \in S(\mathcal{H}_{X0} \otimes \mathcal{H}_{Y0})$; (iii) a map $\mathcal{E}_m : \mathcal{B}(\mathcal{H}_{X0}) \to \mathcal{B}(\mathcal{H}_X)$, such that there are N states $\rho_m = (\mathcal{E} \circ \mathcal{E}_m \otimes \mathbb{1}_{Y0})\omega$ which are mutually adjacent.
Generalized entanglement assisted zero-error capacity	$\hat{C}_E^{(0)}(S) = \lim_{n\to\infty} \frac{1}{n} \log \hat{\alpha}(S^{\otimes n})$	$\hat{\alpha}$ denotes the higher independence number assisted by generalized entanglement which demands that $\mathcal{E}_m(\sigma) = \sum_j E_{jm} \sigma E_{jm}^\dagger$ and that $\sum_j E_{jm}^\dagger E_{jm} \in GL(\mathcal{H}_{X0})$ to be invertible.

where $S^\perp$ is an orthogonal subspace to S given in (8.107). The expression in (8.118) has some relations with the corresponding independence number of the graph. In Example 8.6, for instance, we have that $S^\perp = \varnothing$. It implies that the classical zero-error capacity of the quantum channel $\mathcal{E}$ is equal to zero.

For every confusability graph $S \leq \mathcal{B}(\mathcal{H}_X)$ we have the following relation

$$\alpha_q(S) \leq \alpha(S) \leq \tilde{\alpha}_U(S) \leq \tilde{\alpha}(S) \leq \hat{\alpha}(S) \tag{8.119}$$

where each of these αs, called *independence numbers*, has relation with a different kind of zero-error capacity, as shown in Table 8.2. Detailed information about how to obtain such numbers can be found in the work of Duan et al. [17, 18].

The classical zero-error capacity $C^{(0)}$ and the quantum zero-error capacity $Q^{(0)}$ of a quantum channel were deeply discussed in Chap. 5. The zero-error capacity assisted by entanglement was introduced in the previous section.

Among the independence numbers showed, the only which is not directly related to a zero-error capacity is $\tilde{\alpha}_U(S)$ because it considers unitary restrictions in its definitions. The numbers $\alpha_q(S), \alpha(S), \tilde{\alpha}(S)$ are $\hat{\alpha}(S)$ computable. However, finding a computable expression to the associated zero-error capacity cannot be a trivial task.

Regarding the Lovász theta function, presented in Sect. 4.3 for the classical scenario, it works as an upper bound for the zero-error capacity of a DMC. It is natural to pursue a quantum version of such definition. It was presented by the authors and is formally defined as follows.

Definition 8.5 (Quantum Lovász Theta Number [17]). Let S be the non-commutative graph of a quantum channel $\mathcal{E}$. The quantum Lovász theta number is given by

$$\tilde{\vartheta}(S) = \sup_n \vartheta(S \otimes \mathcal{L}(\mathbf{C}^n)) \tag{8.120}$$

$$= \sup_n \max\{\|\mathbb{1} + T\| : T \in S^\perp \otimes \mathcal{L}(\mathbf{C}^n), \mathbb{1} + T \geq 0\}, \tag{8.121}$$

where the supremum is over all integers n, and the maximum is taken over Hermitian operators T.

The authors show more results regarding characteristics of this quantum version of the Lovász theta number, such as its monotonicity and supermultiplicativity.

Considering the independence numbers and their corresponding zero-error capacities as well as the Lovász theta number, both characterized using the same approach of non-commutative graphs, the authors believe that these results suggest that there might be a much more systematic way in which operator systems generalize Graph Theory to the non-commutative domain. They are pursuing new results in such direction.

8.6 $\mathcal{QMA}$-Completeness of Quantum Clique

In this section we are going to explore the results of Belgi and Shor regarding the computational complexity of the quantum clique problem which was found out to be $\mathcal{QMA}$-complete [5]. According to the authors, the original problem of finding the quantum clique can be written in terms of finding the zero-error capacity of a quantum channel. Exploring the zero-error behavior in this scenario brought relevant contributions to the theory of complexity, enlightening the classification of an important problem according to quantum complexity classes.

Theory of Complexity is a subarea of computer science whose goal is to prove for important problems that their solutions require certain minimum resources [38]. When considering each solution, it takes into account a *model of computation* (classical Turing machines, probabilistic Turing machines, quantum Turing machines, for instance) and a *certain resource* (for example, memory or time) [34].

Problems are grouped into *complexity classes*, according to the model used and to a function of the amount of resources their best solutions demand over a certain size of input in a worst-case scenario. For classical models of computation and when time resources are considered, the complexity classes $\mathcal{P}$ and $\mathcal{NP}$ are widely studied.

An algorithmic problem belongs to the complexity class $\mathcal{P}$ of *polynomially solvable problems* if it can be solved by an algorithm with polynomial worst-case runtime [38]. The class $\mathcal{P}$ is felt to capture the notion of problems with efficient time solutions considering classical (deterministic) Turing machines [2].

The complexity class $\mathcal{NP}$ contains problems having efficiently verifiable solutions. In other words, if x is a solution to the problem (certificate), it is possible to verify that in deterministic polynomial time [2]. For example, the subset sum is in the $\mathcal{NP}$ class. In this problem, given a list of n numbers $A_1, A_2, \ldots, A_n$ and a number T, one must decide if there is a subset of numbers that sums up to T. The certificate for this problem is the list of members in this subset. For a practical example in the problem considered, if the list of numbers is $\{4, -8, 0, 22, -17, 3, 2\}$ and $T = 1$, then the certificate $\{4, -8, 3, 2\}$ can be verified as a solution to the problem in polynomial time.

Any problem in $\mathcal{P}$ is also in the $\mathcal{NP}$ class because we can solve it in polynomial time even without the need of a certificate [12]. However, the question if the classes $\mathcal{P}$ and $\mathcal{NP}$ are equal or different remains as one of the most important challenges for computer science. Certain problems in the $\mathcal{NP}$ class, in particular, have a special classification. A problem in $\mathcal{NP}$ is called an $\mathcal{NP}$-complete problem if any efficient algorithm for it can be converted into an efficient algorithm for any other problem in $\mathcal{NP}$ [20].

Considering this brief introduction of computational complexity, from now on we will examine more closely the contribution and the results of Beigi and Shor [5]. Some definitions and argumentation presented below in this section are from their original work and the reader is referred to it for more details.

The *clique of a graph* is a widely known $\mathcal{NP}$-complete problem. Given a graph G a clique is a subset of vertices, every two of which are adjacent, and the size of a clique is the number of its vertices. The clique problem is that given a graph G and an integer number k, decide whether G contains a clique of size k or not.

Let G^C be the complement of the graph G. In the complement of G a clique is changed to an *independent set*. A subset of a graph G where no two vertices are adjacent characterize an independent set. The maximum size of an independence set is the *independence number* of the graph G, denoted by $\alpha(G)$. So the clique problem in the complement graph reduces to decide whether $\alpha(G) \geq k$, and then it is $\mathcal{NP}$-complete. This reduction is important because the problem of computing $\alpha(G)$ is related to the problem of computing the zero-error capacity of a classical discrete memoryless channel, as shown extensively on Chap. 4.

The quantum version of the clique problem, known as *quantum clique problem*, is also to decide whether $\alpha(\mathcal{E}) \geq k$ for a given quantum channel $\mathcal{E}$. It is equivalent to decide whether there exists quantum states $\rho_1, \ldots, \rho_k$ such that $\mathcal{E}(\rho_1), \ldots, \mathcal{E}(\rho_k)$ have orthogonal supports or not. Note that, for any two states σ_1, σ_2, then $\mathrm{Tr}(\sigma_1 \sigma_2) \geq 0$ and equality holds iff σ_1 and σ_2 have orthogonal supports.

Let $\sigma_{1,2} = \sigma_1 \otimes \sigma_2$ then $\mathrm{Tr}(\sigma_1 \sigma_2) = \mathrm{Tr}(S\sigma_{1,2})$, where S is the swap gate $(S |\psi\rangle |\varphi\rangle = |\varphi\rangle |\psi\rangle)$. We can estimate $\mathrm{Tr}(\sigma_1 \sigma_2)$ by applying the swap gate. We must notice that if $\sigma_{1,2}$ is not separable then the equality does not hold and the orthogonality is not implied by $\mathrm{Tr}(S\sigma_{1,2}) = 0$. To avoid this problem we must restrict ourselves to entanglement breaking channels.

Definition 8.6 (Entanglement Breaking Quantum Channel [5]). A quantum channel $\mathcal{E}$ is called an entanglement breaking quantum channel if there are POVM

$\{M_i\}$ and states σ_i such that

$$\mathcal{E}(\rho) = \sum_i \text{Tr}(M_i\rho)\sigma_i \tag{8.122}$$

for any ρ. In this case, $\mathcal{E}^{\otimes 2}(\rho_{1,2})$ is always separable, $\text{Tr}(S\mathcal{E}^{\otimes 2}(\rho_{1,2})) \geq 0$ and equality implies $\mathcal{E}(\rho_1)$ and $\mathcal{E}(\rho_2)$ are orthogonal.

Putting all the concepts together, we can now formally define the quantum clique problem.

Definition 8.7 (Quantum Clique Problem [5]). The quantum clique problem $(\mathcal{E}, k, a, b)$ is defined as follows:

- **Input**. Integer numbers n and k; non-negative real numbers a and b with an inverse polynomial gap $b - a > n^{-c}$; and $\mathcal{E}$ an entanglement breaking quantum channel that acts on n-qubit states;
- **Promise**. Either exists $\rho_1 \otimes \ldots \otimes \rho_k$ such that $\sum_{i,j} \text{Tr}(S\mathcal{E}(\rho_i) \otimes \mathcal{E}(\rho_j)) \leq a$ or for any states $\rho_{1,2,\ldots,k}$ we have $\sum_{i,j} \text{Tr}(S\mathcal{E}^{\otimes 2}(\rho_{i,j})) \geq b$;
- **Output**. Decide which one is the case.

Despite the deep understanding of the clique problem as an $\mathcal{NP}$-complete problem in the classical case, the same does not happen to the quantum clique problem prior to the work discussed here. Even nowadays, our knowledge regarding quantum complexity theory is still not rich as its classical analogue.

Now we are going to characterize the $\mathcal{QMA}$ quantum complexity class that, loosely speaking, is the quantum version of the $\mathcal{NP}$ class [37]. The acronym of this complexity class stands for *Quantum-Merlin-Arthur* where Merlin is an oracle with infinite computational power and Arthur is a quantum polynomial time verifier. Merlin answers decision problems of the type *"Is x in L?"* and accompany the answer with a polynomial certificate y which Arthur can verify in polynomial time using a quantum machine [1]. We associate two probabilities with the $\mathcal{QMA}$ class which are related to the completeness and soundness. The formal definition of such complexity class is presented as follows.

Definition 8.8 ($\mathcal{QMA}$ Complexity Class [37]). A language L is said to be in $\mathcal{QMA}(2/3, 1/3)$ if there exists a quantum polynomial time verifier V such that

- **Completeness**. $\forall x \in L, \exists |\xi\rangle \in \mathcal{H}^{p(|x|)}, \Pr(V(|x\rangle |\xi\rangle) = 1) \geq 2/3$;
- **Soundness**. $\forall x \notin L, \forall |\xi\rangle \in \mathcal{H}^{p(|x|)}, \Pr(V(|x\rangle |\xi\rangle) = 1) \leq 1/3$;

After the $\mathcal{QMA}$ complexity class was characterized, Kitaev introduced a problem from Physics called the *"local Hamiltonian problem"* and showed that it is $\mathcal{QMA}$-complete. This problem is the quantum analogue of the classical SAT problem and these results are the analogue of the Cook-Levin theorem [5]. The formal definition of this problem is described below.

Definition 8.9 (Local Hamiltonian Problem [24]). The k-local Hamiltonian problem $(H_1, \ldots, H_s, a, b)$ is defined as follows:

- **Input.** An integer n, real numbers a, b such that $b - a > n^{-c}$, and polynomially many Hermitian non-negative semidefinite matrices $H_1, \ldots, H_s$ with bounded norm $\|H_i\| \leq 1$, such that each of them acts just on k of n qubits;
- **Promise.** The smallest eigenvalue of $H_1 + \ldots + H_s$ is either less than a or greater than b;
- **Output.** Decide which one is the case.

Intuitively, a k-local matrix assigns a real number to any quantum state on n qubits. This number depends only on the reduced state of the k qubits where a quantum operator M acts non-trivially, and can be thought of as a locally defined penalty on a given quantum state. Loosely speaking, the k-local Hamiltonian problem asks whether there exists a quantum state that can significantly avoid a collection of such penalties [37].

Considering the quantum clique problem, the following theorem states its complexity.

Theorem 8.6 (Quantum Clique is $\mathcal{QMA}$-Complete [5]). *The quantum clique problem $(\mathcal{E}, k, a, b)$ where $\mathcal{E}$ is an entanglement breaking channel on n-qubit states and has the operator-sum representation*

$$\mathcal{E}(\rho) = \sum_{i=1}^{r} E_i \rho E_i^{\dagger}, \tag{8.123}$$

where $\sum_i E_i^{\dagger} E_i = \mathbb{1}$ and $r = poly(n)$, is $\mathcal{QMA}$-complete.

The proof of this theorem consists in showing that $(\mathcal{E}, k, a, b)$ is $\mathcal{QMA}$. To prove the hardness, the authors establish a polynomial time reduction from the local Hamiltonian problem to quantum clique. In this result, a is a positive number that means that some probability of error is allowed.

If we consider the case $a = 0$, we will try to find a protocol with no error. In this case, $(\mathcal{E}, k, a = 0, b)$ exactly says that whether $\alpha(\mathcal{E}) \geq k$ or not. We can achieve this by using a *zero-error quantum channel* $\mathcal{E}$ where

$$\mathcal{E}(\rho) = \sum_{i-1}^{r} \mathrm{Tr}(M_i \rho) |i\rangle \langle i|, \tag{8.124}$$

where $\{M_1, \ldots, M_r\}$ is a POVM and $|1\rangle, \ldots, |r\rangle$ are orthogonal states. Checking orthogonality of two outcomes states is accomplished in the following way: given two states $\mathcal{E}(\rho)$ and $\mathcal{E}(\rho')$ we measure them in the basis $|1\rangle, \ldots, |r\rangle$. If the outcome of the measurements are the same, then their supports are not orthogonal.

So, the quantum clique problem $(\mathcal{E}, k, a = 0, b)$ where $\mathcal{E}$ is a zero-error quantum channel that can be implemented exactly by a polynomial time verifier is $\mathcal{QMA}_1$-complete. The article of Beigi and Shor contains the entire version of the proofs briefly discussed here [5].

Some promise problems in the literature are known to be $\mathcal{QMA}$-complete, such as variants of the local Hamiltonian problem, the density matrix consistency problem, and also other problems about quantum circuits [37]. However, the quantum clique problem is the only so far in this complexity class whose complexity was described using zero-error quantum channels.

This section described the complexity classification of the quantum clique problem. We presented the results of Beigi and Shor that showed a non-trivial result where zero-error quantum channels helped in determining the quantum complexity of a problem. The contributions of these authors enrich the knowledge regarding quantum complexity classes and the classification of an important quantum problem, which may have implications in algorithms and protocols for practical applications.

8.7 Further Reading

In this section we could see some developments in the literature that provide new results and insights into the quantum zero-error information theory. We saw the background which relates quantum zero-error information theory with Kochen-Specker sets and Bell's inequality. The quantum version of the Wielandt's inequality which states an upper bound to the number of uses of a quantum channel in order to map an arbitrary density operator to a full rank operator was also discussed. An alternative version of the zero-error capacity of quantum channels considering entanglement assistance was introduced. The approach for zero-error capacity considering non-commutative graphs and the quantum counterpart of the Lovász theta function was also considered. Lastly, an application of quantum zero-error channels to find the complexity class of a problem showed a non-trivial application of the concepts discussed along the book.

Other recent results besides those discussed here can also be found in the literature. Blume-Kohout et al. [7] developed a framework to handle quantum information that can be perfectly preserved (i.e, with zero-error) by the system dynamics. According to the authors, the system dynamics affects the kind of information that can be carried or store (classical, quantum or neither, for instance). Taking that into account, the main purpose of their operational framework is to describe how to perfectly preserve information despite the system dynamics. This framework considers not only quantum channels with positive zero-error capacity, but also quantum error-correcting codes, decoherence-free subspaces and subsystems and even other methods proposed by the own authors, such as the unconditionally preserved codes. This work provides an exhaustive classification of ways that information can be preserved.

Regarding practical implementations, Gyongyosi and Imre [21] considered the use of multiple optical channels to send information. Each of these individual channels has no positive zero-error capacity, but when used jointly the zero-error capacity is superactivated. Their idea is to adopt such strategy as part of the implementation of quantum repeaters, devices that can extend the range of quantum communication between sender and receiver.

Besides the already described results on superactivation of zero-error capacity, Shirokov [33] showed a special kind of superactivation of quantum channels under block coding.

Briё t et al. considered the use of quantum entanglement in the zero-error source-channel coding problem [8]. In their scenario, Alice and Bob are each given an input from a random source and get access to a noisy channel through which Alice can send messages to Bob. Their goal is to minimize the average number of channel uses per source input such that Bob can learn Alice's inputs with zero probability of error. Their results show lower bound and optimum rate of entanglement-assisted source codes and the advantage that entanglement can give in the source-channel coding problem.

We hope that much more results on quantum zero-error information theory are yet to come.

References

1. Aharonov D, Naveh T (2002) Quantum NP – a survey. http://arxiv.org/abs/quant-ph/0210077. Accessed 02 Feb 2016
2. Arora S, Barak B (2009) Computational complexity: a modern approach. Cambridge University Press, Cambridge
3. Aspect A, Grangier P, Roger G (1982) Experimental realization of Einstein-Podolsky-Rosen-Bohm Gedankenexperiment: a new violation of Bell's inequalities. Phys Rev Lett 49(2):91–94
4. Beigi S (2010) Entanglement-assisted zero-error capacity is upper bounded by the lovász theta function. Phys Rev A 82:010303
5. Beigi S, Shor PW (2008) On the complexity of computing zero-error and Holevo capacity of quantum channels. http://arxiv.org/abs/0709.2090. Accessed 25 Oct 2014
6. Bengtsson I, Zyczkowski K (2006) Geometry of quantum states. Cambridge University Press, Cambridge
7. Blume-Kohout R, Ng HK, Poulin D, Viola L (2010) Information-preserving structures: A general framework for quantum zero-error information. Phys Rev A 82:062306
8. Briet JB, Buhrman H, Laurent M, Piovesan T, Scarpa G (2015) Entanglement-assisted zero-error source-channel coding. IEEE Trans Inf Theory 61(2):1124–1138
9. Cabello A, Badziag P, Cunha MT, Bourennane M (2013) Simple hardy-like proof of quantum contextuality. Phys Rev Lett 111:180404
10. Cameron PJ, Montanaro A, Newman MW, Severini S, Winter A (2007) On the quantum chromatic number of a graph. Electron J Comb 81(14):1–15
11. Clauser J, Horne MA, Shimony A, Holt RA (1969) Proposed experiment to test local hidden-variable theories. Phys Rev Lett 41:880
12. Cormen TH, Leiserson CE, Rivest RL, Stein C (2009) Introduction to algorithms, 3rd edn. MIT, Cambridge
13. Cubitt T, Leung D, Matthews W, Winter A (2010) Improving zero-error classical communication with entanglement. Phys Rev Lett 104:230503
14. Cubitt TS, Leung D, Matthews W, Winter A (2010) Improving zero-error classical communication with entanglement. Phys Rev Lett 104
15. Cubitt TS, Leung D, Matthews W, Winter A (2011) Zero-error channel capacity and simulation assisted by non-local correlations. IEEE Trans Inf Theory 57(8):5509–5523
16. D'Ambrosio V, Herbauts I, Anselem E, Nagah E (2013) Experimental implementation of a Kochen-Sepecker set of quantum tests. Phys Rev X 3:011012

17. Duan R, Severini S, Winter A (2011) Zero-error communication via quantum channels, non-commutative graphs and a quantum Lovasz ϑ function. In: IEEE international symposium on information theory, Russia, pp 64–68
18. Duan R, Severini S, Winter A (2013) Zero-error communication via quantum channels, non-commutative graphs and a quantum Lovasz ϑ function. IEEE Trans Inf Theory 59(2):1164–1174
19. Einstein A, Poldosky B, Rosen N (1935) Can quantum-mechanical description of physical reality be considered complete? Phys Rev 47:777–780
20. Goldreich O (2010) P, NP, and NP completeness. Cambridge University Press, Cambridge
21. Gyongyosi L, Imre S (2012) Long-distance quantum communications with superactivated gaussian optical quantum channels. Opt Eng 51(1):1–16
22. Horn RA, Johnson CR (1985) Matrix analysis. Cambridge University Press, Cambridge
23. Imre S, Gyongyosi L (2012) Advanced quantum communications: an engineering approach, 1st edn. Wiley, New York
24. Kitaev A, Shen A, Vyalyi MN (2002) Classical and quantum computation. American Mathematical Society, Rhode Island
25. Leung D, Mancinska L, Matthews W, Ozols M, Roy A (2012) Entanglement can increase asymptotic rates of zero-error classical communication over classical channels. Commun Math Phys 311:97–111
26. Mancinska L, Scarpa G, Severini S (2013) A generalization of Kochen-Specker sets relates quantum coloring to entanglement-assisted channel capacity. IEEE Trans Inf Theory 59(6):4025–4032
27. Mermin ND (1993) Hidden variables and the two theorems of John Bell. Rev Mod Phys 65(3):803–815
28. Papoulis A, Pillai SU (2002) Probability, random variables and stochastic processes. McGraw-Hill, New York
29. Peres A (1995) Quantum theory: concepts and methods. Kluwer Academic Publishers, Dordrecht
30. Petri CA (1996) Nets, time and space. Theor Comput Sci 3:3–48
31. Sanz M, Pérez-García D, Woff MM, Cirac JI (2010) A quantum version of Wielandt's inequality. IEEE Trans Inf Theory 56(9):4668–4673
32. Scarpa G, Severini S (2012) Kochen-Specker sets and the rank-1 quantum chromatic number. IEEE Trans Inf Theory 58(4):2524–2529
33. Shirokov ME (2015) On quantum zero-error capacity. Commun Mosc Math Soc 70(1):176–178
34. Sipser M (2012) Introduction to the theory of computation, 3rd edn. Course Technology, Boston
35. van Dam W, Gill RD, Grünwald PD (2005) The statistical strength of nonlocality proofs. IEEE Trans Inf Theory 51(8):2812–2835
36. von Neumann J (1955) Mathematical foundations of quantum mechanics. Princeton University Press, Princeton
37. Watrous J (2008) Quantum computational complexity. http://arxiv.org/abs/0804.3401. Accessed 02 Feb 2016
38. Wegener I (2005) Complexity theory – exploring the limits of efficient algorithms. Springer, Dortmund
39. Young HD (1964) Fundamentals of mechanics and heat, 1st edn. McGraw-Hill Book Company, Inc., New York

Index

Symbols
†-Algebra, 110

A
Adjacency, 69, 71, 72, 82, 91, 113, 123, 170, 172
  between classical symbols, **67**
  of quantum states, **82**
  reducing mapping, **69**, 73
Algebra
  interaction, 110
Amplitude, 9, 12
  damping, 116, 127

B
Basis
  computational, 8, 86
  extendible, **97**
  Hadamard, 14, 15
  orthogonal, 170
  orthonormal, 15, 49, 87, 162
  unextendible, **97**
Bell
  inequality, 150, 153
  labs, 28
  states, **22**, 44, 90
Bit, **8**, 10, 29, 93
Bloch sphere, 10, 145
Bra, **9**

C
Capacity
  $C_{1,1}$, **53**

$C_{1,A}$, 58
  adaptive, 58
  classical, 37
  entanglement-assisted, **58**, 58, 169
  HSW, **55**, 55, 80, 87, 121
  one-shot, 79, 89
  one-shot zero-error, 93, 98, 104
  ordinary, **37**, 65, 66, 68, 77, 87, 121
  quantum, 53, 58, 59, **60**, 60, 130
  quantum secrecy, 116
  secrecy, **115**, 121, 129, 130
  secrecy symmetric, 116
  Shannon capacity, **71**
  zero-error, 4, 50, 65, 66, **68**, 71, 73, 77, 79, **81**, 81, 96, 100, 117, 121, 140, 141, 160, 172, 175, 176
    activation, 92
    entanglement-assisted, **171**
    zero-error asymptotic, 100
    zero-error secrecy, 107, 120, 130
Cartesian product, 77, 163
Chain rule, **32**, **45**
Channel, 3
  G_5, 68
  binary erasure, 38
  binary symmetric, 36, 138
  bit-flip, 38, 48, 56
  capacity, 28, 54
  Choi-Jamiołkowski isomorphism, 49
  classical, **34**, 34, 141, 171
  coding theorem, 1, **40**, 40, 55, 66
  composite, 94, 104
  confusability graph of, 172
  covariant, 130
  depolarizing, 49, 55

© Springer International Publishing Switzerland 2016
E.B. Guedes et al., *Quantum Zero-Error Information Theory*,
DOI 10.1007/978-3-319-42794-2

Printed in the United States
By Bookmasters